GOVERNORS STATE UNIVERSITY LIBRARY

3 1611 00333 1458

D1711755

Solar Energy Fundamentals
and Modeling Techniques

Zekai Şen

Solar Energy Fundamentals and Modeling Techniques

Atmosphere, Environment, Climate Change and Renewable Energy

 Springer

Prof. Zekai Şen
İstanbul Technical University
Faculty of Aeronautics and Astronautics
Dept. Meteorology
Campus Ayazaga
34469 İstanbul
Turkey

TJ
810
.S454
2008

ISBN 978-1-84800-133-6   e-ISBN 978-1-84800-134-3

DOI 10.1007/978-1-84800-134-3

British Library Cataloguing in Publication Data
Sen, Zekai
   Solar energy fundamentals and modeling techniques :
   atmosphere, environment, climate change and renewable
   energy
   1. Solar energy
   I. Title
   621.4'7
ISBN-13: 9781848001336

Library of Congress Control Number: 2008923780

© 2008 Springer-Verlag London Limited

Apart from any fair dealing for the purposes of research or private study, or criticism or review, as permitted under the Copyright, Designs and Patents Act 1988, this publication may only be reproduced, stored or transmitted, in any form or by any means, with the prior permission in writing of the publishers, or in the case of reprographic reproduction in accordance with the terms of licences issued by the Copyright Licensing Agency. Enquiries concerning reproduction outside those terms should be sent to the publishers.

The use of registered names, trademarks, etc. in this publication does not imply, even in the absence of a specific statement, that such names are exempt from the relevant laws and regulations and therefore free for general use.

The publisher makes no representation, express or implied, with regard to the accuracy of the information contained in this book and cannot accept any legal responsibility or liability for any errors or omissions that may be made.

Cover design: eStudio Calamar S.L., Girona, Spain

Printed on acid-free paper

9 8 7 6 5 4 3 2 1

springer.com

Bismillahirrahmanirrahim

In the name of Allah the most merciful and the most beneficial

Preface

Atmospheric and environmental pollution as a result of extensive fossil fuel exploitation in almost all human activities has led to some undesirable phenomena that have not been experienced before in known human history. They are varied and include global warming, the greenhouse affect, climate change, ozone layer depletion, and acid rain. Since 1970 it has been understood scientifically by experiments and research that these phenomena are closely related to fossil fuel uses because they emit greenhouse gases such as carbon dioxide (CO_2) and methane (CH_4) which hinder the long-wave terrestrial radiation from escaping into space and, consequently, the earth troposphere becomes warmer. In order to avoid further impacts of these phenomena, the two main alternatives are either to improve the fossil fuel quality thus reducing their harmful emissions into the atmosphere or, more significantly, to replace fossil fuel usage as much as possible with environmentally friendly, clean, and renewable energy sources. Among these sources, solar energy comes at the top of the list due to its abundance and more even distribution in nature than other types of renewable energy such as wind, geothermal, hydropower, biomass, wave, and tidal energy sources. It must be the main and common purpose of humanity to develop a sustainable environment for future generations. In the long run, the known limits of fossil fuels compel the societies of the world to work jointly for their replacement gradually by renewable energies rather than by improving the quality of fossil sources.

Solar radiation is an integral part of different renewable energy resources, in general, and, in particular, it is the main and continuous input variable from the practically inexhaustible sun. Solar energy is expected to play a very significant role in the future especially in developing countries, but it also has potential in developed countries. The material presented in this book has been chosen to provide a comprehensive account of solar energy modeling methods. For this purpose, explanatory background material has been introduced with the intention that engineers and scientists can benefit from introductory preliminaries on the subject both from application and research points of view.

The main purpose of Chapter 1 is to present the relationship of energy sources to various human activities on social, economic and other aspects. The atmospheric environment and renewable energy aspects are covered in Chapter 2. Chapter 3 provides the basic astronomical variables, their definitions and uses in the calculation of the solar radiation (energy) assessment. These basic concepts, definitions, and derived astronomical equations furnish the foundations of the solar energy evaluation at any given location. Chapter 4 provides first the fundamental assumptions in the classic linear models with several modern alternatives. After the general review of available classic non-linear models, additional innovative non-linear models are presented in Chapter 5 with fundamental differences and distinctions. Fuzzy logic and genetic algorithm approaches are presented for the non-linear modeling of solar radiation from sunshine duration data. The main purpose of Chapter 6 is to present and develop regional models for any desired location from solar radiation measurement sites. The use of the geometric functions, inverse distance, inverse distance square, semivariogram, and cumulative semivariogram techniques are presented for solar radiation spatial estimation. Finally, Chapter 7 gives a summary of solar energy devices.

Applications of solar energy in terms of low- and high-temperature collectors are given with future research directions. Furthermore, photovoltaic devices are discussed for future electricity generation based on solar power site-exploitation and transmission by different means over long distances, such as fiber-optic cables. Another future use of solar energy is its combination with water and, as a consequence, electrolytic generation of hydrogen gas is expected to be another source of clean energy. The combination of solar energy and water for hydrogen gas production is called solar-hydrogen energy. Necessary research potentials and application possibilities are presented with sufficient background. New methodologies that are bound to be used in the future are mentioned and, finally, recommendations and suggestions for future research and application are presented, all with relevant literature reviews. I could not have completed this work without the support, patience, and assistance of my wife Fatma Şen.

İstanbul, Çubuklu

15 October 2007

Contents

1 **Energy and Climate Change** 1
   1.1 General 1
   1.2 Energy and Climate 3
   1.3 Energy and Society 5
   1.4 Energy and Industry 10
   1.5 Energy and the Economy 12
   1.6 Energy and the Atmospheric Environment 13
   1.7 Energy and the Future 17
   References 18

2 **Atmospheric Environment and Renewable Energy** 21
   2.1 General 21
   2.2 Weather, Climate, and Climate Change 22
   2.3 Atmosphere and Its Natural Composition 26
   2.4 Anthropogenic Composition of the Atmosphere 28
      2.4.1 Carbon Dioxide (CO_2) 29
      2.4.2 Methane (CH_4) 30
      2.4.3 Nitrous Oxide (N_2O) 31
      2.4.4 Chlorofluorocarbons (CFCs) 31
      2.4.5 Water Vapor (H_2O) 31
      2.4.6 Aerosols 33
   2.5 Energy Dynamics in the Atmosphere 34
   2.6 Renewable Energy Alternatives and Climate Change 35
      2.6.1 Solar Energy 36
      2.6.2 Wind Energy 37
      2.6.3 Hydropower Energy 38
      2.6.4 Biomass Energy 39
      2.6.5 Wave Energy 40
      2.6.6 Hydrogen Energy 41
   2.7 Energy Units 43
   References 44

3 Solar Radiation Deterministic Models 47
- 3.1 General 47
- 3.2 The Sun 47
- 3.3 Electromagnetic (EM) Spectrum 51
- 3.4 Energy Balance of the Earth 55
- 3.5 Earth Motion 57
- 3.6 Solar Radiation 61
  - 3.6.1 Irradiation Path 64
- 3.7 Solar Constant 66
- 3.8 Solar Radiation Calculation 67
  - 3.8.1 Estimation of Clear-Sky Radiation 70
- 3.9 Solar Parameters 72
  - 3.9.1 Earth's Eccentricity 72
  - 3.9.2 Solar Time 72
  - 3.9.3 Useful Angles 74
- 3.10 Solar Geometry 77
  - 3.10.1 Cartesian and Spherical Coordinate System 78
- 3.11 Zenith Angle Calculation 85
- 3.12 Solar Energy Calculations 87
  - 3.12.1 Daily Solar Energy on a Horizontal Surface 88
  - 3.12.2 Solar Energy on an Inclined Surface 91
  - 3.12.3 Sunrise and Sunset Hour Angles 93
- References 98

4 Linear Solar Energy Models 101
- 4.1 General 101
- 4.2 Solar Radiation and Daylight Measurement 102
  - 4.2.1 Instrument Error and Uncertainty 103
  - 4.2.2 Operational Errors 104
  - 4.2.3 Diffuse-Irradiance Data Measurement Errors 105
- 4.3 Statistical Evaluation of Models 106
  - 4.3.1 Coefficient of Determination (R^2) 109
  - 4.3.2 Coefficient of Correlation (r) 110
  - 4.3.3 Mean Bias Error, Mean of Absolute Deviations, and Root Mean Square Error 111
  - 4.3.4 Outlier Analysis 112
- 4.4 Linear Model 113
  - 4.4.1 Ångström Model (AM) 116
- 4.5 Successive Substitution (SS) Model 120
- 4.6 Unrestricted Model (UM) 126
- 4.7 Principal Component Analysis (PCA) Model 133
- 4.8 Linear Cluster Method (LCM) 140
- References 147

Contents   xi

5 Non-Linear Solar Energy Models ... 151
- 5.1   General ... 151
- 5.2   Classic Non-Linear Models ... 151
- 5.3   Simple Power Model (SPM) ... 156
  - 5.3.1   Estimation of Model Parameters ... 157
- 5.4   Comparison of Different Models ... 159
- 5.5   Solar Irradiance Polygon Model (SIPM) ... 160
- 5.6   Triple Solar Irradiation Model (TSIM) ... 168
- 5.7   Triple Drought–Solar Irradiation Model (TDSIM) ... 172
- 5.8   Fuzzy Logic Model (FLM) ... 176
  - 5.8.1   Fuzzy Sets and Logic ... 177
  - 5.8.2   Fuzzy Algorithm Application for Solar Radiation ... 179
- 5.9   Geno-Fuzzy Model (GFM) ... 186
- 5.10 Monthly Principal Component Model (MPCM) ... 188
- 5.11 Parabolic Monthly Irradiation Model (PMIM) ... 196
- 5.12 Solar Radiation Estimation from Ambient Air Temperature ... 202
- References ... 206

6 Spatial Solar Energy Models ... 209
- 6.1   General ... 209
- 6.2   Spatial Variability ... 210
- 6.3   Linear Interpolation ... 212
- 6.4   Geometric Weighting Function ... 214
- 6.5   Cumulative Semivariogram (CSV) and Weighting Function ... 216
  - 6.5.1   Standard Spatial Dependence Function (SDF) ... 217
- 6.6   Regional Estimation ... 220
  - 6.6.1   Cross-Validation ... 221
  - 6.6.2   Spatial Interpolation ... 226
- 6.7   General Application ... 228
- References ... 236

7 Solar Radiation Devices and Collectors ... 239
- 7.1   General ... 239
- 7.2   Solar Energy Alternatives ... 239
- 7.3   Heat Transfer and Losses ... 241
  - 7.3.1   Conduction ... 242
  - 7.3.2   Convection ... 243
  - 7.3.3   Radiation ... 244
- 7.4   Collectors ... 245
  - 7.4.1   Flat Plate Collectors ... 246
  - 7.4.2   Tracking Collectors ... 249
  - 7.4.3   Focusing (Concentrating) Collectors ... 250
  - 7.4.4   Tilted Collectors ... 252
  - 7.4.5   Solar Pond Collectors ... 253
  - 7.4.6   Photo-Optical Collectors ... 253

	7.5	Photovoltaic (PV) Cells	256
	7.6	Fuel Cells	259
	7.7	Hydrogen Storage and Transport	259
	7.8	Solar Energy Home	260
	7.9	Solar Energy and Desalination Plants	261
	7.10	Future Expectations	262
	References	264
A	**A Simple Explanation of Beta Distribution**	267
B	**A Simple Power Model**	269
Index		273

Chapter 1
Energy and Climate Change

1.1 General

Energy and fresh water are the two major commodities that furnish the fundamentals of every human activity for a reasonable and sustainable quality of life. Energy is the fuel for growth, an essential requirement for economic and social development. Solar energy is the most ancient source and the root for almost all fossil and renewable types. Special devices have been used for benefiting from the solar and other renewable energy types since time immemorial. During the early civilizations water and wind power have been employed as the major energy sources for navigation, trade, and information dissemination. For instance, Ebul-İz Al-Jazari (1136–1206), as mentioned by Şen (2005), was the first scientist who developed various instruments for efficient energy use. Al-Jazari described the first reciprocating piston engine, suction pump, and valve, when he invented a two-cylinder reciprocating suction piston pump, which seems to have had a direct significance in the development of modern engineering. This pump is driven by a water wheel (water energy) that drives, through a system of gears, an oscillating slot-rod to which the rods of two pistons are attached. The pistons work in horizontally opposed cylinders, each provided with valve-operated suction and delivery pipes. His original drawing in Fig. 1.1a shows the haulage of water by using pistons, cylinders, and a crank moved by panels subject to wind power. In Fig. 1.1b the equivalent instrument design is achieved by Hill (1974).

Ebul-İz Al-Jazari's original *robotic* drawing is presented in Fig. 1.2. It works with water power through right and left nozzles, as in the figure, and accordingly the right and left hands of the human figure on the elephant move up and down.

In recent centuries the types and magnitudes of the energy requirements have increased in an unprecedented manner and mankind seeks for additional energy sources. Today, energy is a continuous driving power for future social and technological developments. Energy sources are vital and essential ingredients for all human transactions and without them human activity of all kinds and aspects cannot be progressive. Population growth at the present average rate of 2% also exerts extra pressure on limited energy sources.

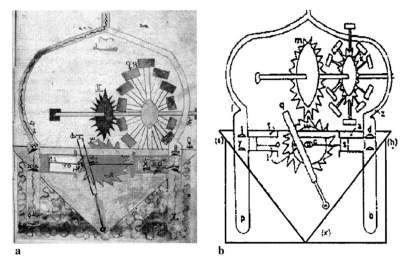

Fig. 1.1 **a** Al-Jazari (1050). **b** Hill (1974)

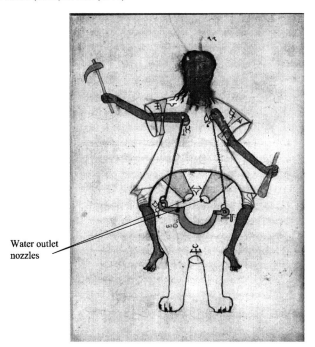

Water outlet nozzles

Fig. 1.2 Robotic from Al-Jazari

The *oil crises* of the 1970s have led to a surge in research and development of renewable and especially solar energy alternatives. These efforts were strongly correlated with the fluctuating market price of energy and suffered a serious setback as this price later plunged. The missing ingredient in such a process was a long-term perspective that hindered the research and development policy within the wider context of fossil and solar energy tradeoffs rather than reactions to temporary price fluctuations. The same events also gave rise to a rich literature on the optimal exploitation of natural resources, desirable rate of research, and development efforts to promote competitive technologies (Tsur and Zemel 1998). There is also a vast amount of literature on energy management in the light of atmospheric pollution and climate change processes (Clarke 1988; Edmonds and Reilly 1985, 1993; Hoel and Kvendokk 1996; Nordhaus 1993, 1997; Tsur and Zemel 1996; Weyant 1993).

The main purpose of this chapter is to present the relationship of energy sources to various human activities including social, economic, and other aspects.

1.2 Energy and Climate

In the past, natural weather events and climate phenomena were not considered to be interrelated with the energy sources, however during the last three decades their close interactions become obvious in the atmospheric composition, which drives the meteorological and climatologic phenomena. Fossil fuel use in the last 100 years has loaded the atmosphere with additional constituents and especially with carbon dioxide (CO_2), the increase of which beyond a certain limit influences atmospheric events (Chap. 2). Since the nineteenth century, through the advent of the industrial revolution, the increased emissions of various *greenhouse* gases (CO_2, CH_4, N_2O, *etc*.) into the atmosphere have raised their concentrations at an alarming rate, causing an abnormal increase in the earth's average temperature. Scientists have confirmed, with a high degree of certainty, that the recent trend in global average temperatures is not a normal phenomenon (Rozenzweig *et al.*, 2007). Its roots are to be found in the unprecedented industrial growth witnessed by the world economy, which is based on energy consumption.

Since climate modification is not possible, human beings must be careful in their use of energy sources and reduce the share of fossil fuels as much as possible by replacing their role with clean and environmentally friendly energy sources that are renewable, such as solar, wind, water, and biomass. In this manner, the extra loads on the atmosphere can be reduced to their natural levels and hence sustainability can be passed on to future generations.

Over the last century, the amount of CO_2 in the atmosphere has risen, driven in large part by the usage of *fossil fuels*, but also by other factors that are related to rising population and increasing consumption, such as land use change, *etc*. On the global scale, increase in the emission rates of greenhouse gases and in particular CO_2 represents a colossal threat to the world climate. Various theories and calculations in atmospheric research circles have already indicated that, over the last half century, there appeared a continuously increasing trend in the average temperature value up to 0.5 °C. If this trend continues in the future, it is expected that in some areas of the world, there will appear extreme events such as excessive rainfall and consequent floods, droughts, and also local imbalances in the natural climatic behavior giving rise to unusual local heat and cold. Such events will also affect the world food production rates. In addition, global temperatures could rise by a further 1 – 3.5 °C by the end of the twenty-first century, which may lead potentially to disruptive climate change in many places. By starting to manage the CO_2 emissions through renewable energy sources now, it may be possible to limit the effects of climate change to adaptable levels. This will require adapting the world's energy systems. Energy policy must help guarantee the future supply of energy and drive the necessary transition. International cooperation on the climate issue is a prerequisite for achieving cost-effective, fair, and sustainable solutions.

At present, the global energy challenge is to tackle the threat of climate change, to meet the rising demand for energy, and to safeguard security of energy supplies. *Renewable energy* and especially solar radiation are effective energy technologies that are ready for global deployment today on a scale that can help tackle climate change problems. Increase in the use of renewable energy reduces CO_2 emissions, cuts local air pollution, creates high-value jobs, curbs growing dependence of one country on imports of fossil energy (which often come from politically unstable regions), and prevents society a being hostage to finite energy resources.

In addition to *demand*-side impacts, *energy production* is also likely to be affected by climate change. Except for the impacts of extreme weather events, research evidence is more limited than for energy consumption, but climate change could affect energy production and supply as a result of the following (Wilbanks *et al.*, 2007):

1. If extreme weather events become more intense
2. If regions dependent on water supplies for hydropower and/or thermal power plant cooling face reductions in water supplies
3. If changed conditions affect facility siting decisions
4. If conditions change (positively or negatively) for biomass, wind power, or solar energyproductions

Climate change is likely to affect both energy use and energy production in many parts of the world. Some of the possible impacts are rather obvious. Where the climate warms due to climate change, less heating will be needed for industrial increase (Cartalis *et al.*, 2001), with changes varying by region and by season. Net energy demand on a national scale, however, will be influenced by the structure of energy supply. The main source of energy for cooling is electricity, while coal, oil, gas, biomass, and electricity are used for space heating. Regions with substantial requirements for both cooling and heating could find that net annual electricity demands increase while demands for other heating energy sources decline (Hadley *et al.*, 2006). Seasonal variation in total energy demand is also important. In some cases, due to infrastructure limitations, peak energy demand could go beyond the maximum capacity of the transmission systems. Tol (2002a,b) estimated the effects of climate change on the demand for global energy, extrapolating from a simple country-specific (UK) model that relates the energy used for heating or cooling to degree days, per capita income, and energy efficiency. According to Tol, by 2100 benefits (reduced heating) will be about 0.75% of gross domestic product

(GDP) and damages (increased cooling) will be approximately 0.45%, although it is possible that migration from heating-intensive to cooling-intensive regions could affect such comparisons in some areas (Wilbanks *et al.*, 2007).

Energy and climate are related concerning cooling during hot weather. Energy use has been and will continue to be affected by climate change, in part because air-conditioning, which is a major energy use particularly in developed countries, is climate-dependent. However, the extent to which temperature rise has affected energy use for space heating/cooling in buildings is uncertain. It is likely that certain adaptation strategies (*e.g.*, tighter building energy standards) have been (or would be) taken in response to climate change. The energy sector can adapt to *climate-change* vulnerabilities and impacts by anticipating possible impacts and taking steps to increase its resilience, *e.g.*, by diversifying energy supply sources, expanding its linkages with other regions, and investing in technological change to further expand its portfolio of options (Hewer 2006). Many energy sector strategies involve high capital costs, and social acceptance of climate-change response alternatives that might imply higher energy prices.

Climate change could have a negative impact on thermal power production since the availability of cooling water may be reduced at some locations because of climate-related decreases (Arnell *et al.*, 2005) or seasonal shifts in river runoff (Zierl and Bugmann 2005). The distribution of energy is also vulnerable to climate change. There is a small increase in line resistance with increasing mean temperatures coupled with negative effects on line sag and gas pipeline compressor efficiency due to higher maximum temperatures. All these combined effects add to the overall uncertainty of climate change impacts on power grids.

1.3 Energy and Society

Since the *energy crisis* in 1973 air pollution from combustion processes has caused serious damage and danger to forests, monuments, and human health in many countries, as has been documented by official studies and yearly statistics. Many environmental damages, including *acid rain* and their forest-damaging consequences, have incurred economic losses in the short term and especially in the long term. Hence, seemingly cheap energy may inflict comparatively very high expenses on society. Figure 1.3 shows three partners in such a social problem including material beneficiary, heat beneficiary, and, in between, the third party who has nothing to do with these two major players.

On the other hand, the climate change due to CO_2 emission into the atmosphere is another example of possible social costs from the use of energy, which is handed over to future generations by today's energy consumers. Again the major source of climate change is the combustion of unsuitable quality fossil fuels.

Today, the scale of development of any society is measured by a few parameters among which the used or the per capita energy amount holds the most significant rank. In fact, most industrialized countries require reliable, efficient, and readily

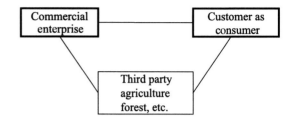

Fig. 1.3 Energy usage partners

available energy for their transportation, industrial, domestic, and military systems. This is particularly true for developing countries, especially those that do not possess reliable and sufficient energy sources.

Although an adequate supply of energy is a prerequisite of any modern society for economic growth, energy is also the main source of environmental and atmospheric pollution (Sect. 1.6). On the global scale, increasing emissions of air pollution are the main causes of greenhouse gases and climate change. If the trend of increasing CO_2 continues at the present rate, then major climatic disruptions and local imbalances in the hydrological as well as atmospheric cycles will be the consequences, which may lead to excessive rainfall or drought, in addition to excessive heat and cold. Such changes are already experienced and will also affect the world's potential for food production. The continued use of conventional energy resources in the future will adversely affect the natural environmental conditions and, consequently, social energy-related problems are expected to increase in the future. A new factor, however, which may alleviate the environmental and social problems of future energy policies, or even solve them, is the emerging new forms of renewable sources such as solar, wind, biomass, small hydro, wave, and geothermal energies, as well as the possibility of solar hydrogen energy.

The two major reasons for the increase in the energy consumption at all times are the steady population increase and the strive for better development and comfort. The world *population* is expected to almost double in the next 50 years, and such an increase in the population will take place mostly in the developing countries, because the developed countries are not expected to show any significant population increase. By 2050, energy demand could double or triple as population rises and developing countries expand their economies and overcome poverty.

The energy demand growth is partially linked to population growth, but may also result from larger per capita energy consumptions. The demand for and production of energy on a world scale are certain to increase in the foreseeable future. Of course, growth will definitely be greater in the developing countries than in the industrialized ones. Figure 1.4 shows the world population increase for a 100-year period with predictions up to 2050. It indicates an exponential growth trend with increasing rates in recent years such that values double with every passage of a fixed amount of time, which is the *doubling time*.

The recent rise in population is even more dramatic when one realizes that per capita consumption of energy is also rising thus compounding the effects. Economic growth and the population increase are the two major forces that will continue to

1.3 Energy and Society

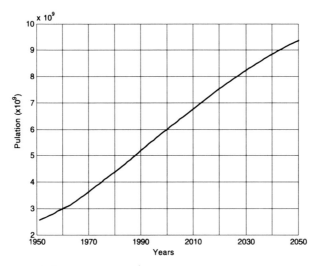

Fig. 1.4 Human population

cause increase in the energy demand during the coming decades. The future energy demand is shown in Table 1.1 for the next 30 years (Palz 1994).

The energy use of a society distinguishes its scale of development compared to others. A poor citizen in a less-developed country must rely on human and animal power. In contrast, developed countries consume large quantities of energy for transportation and industrial uses as well as heating and cooling of building spaces.

How long can the world population want these percentages to increase? The answer is not known with certainty. If the growth rate, G_r, is 1% per year then the doubling period, D_p, will be 69 years. Accordingly, the doubling periods, are presented for different growth rates in Fig. 1.5. It appears as a straight line on double-logarithmic paper, which implies that the model can be expressed mathematically in the form of a power function, as follows:

$$D_p = 69 G_r^{-0.98}. \tag{1.1}$$

It is obvious that there is an inversely proportional relationship between the population growth rate and the doubling period.

Table 1.1 Future energy demand

1000 Moet	1990	2020	Increase (%)
Industrialized countries	4.1	4.6	12
Central and eastern Europe	1.7	1.8	5
Developing countries	2.9	6.9	137
World	8.7	13.3	52

Moet million oil equivalent ton (energy unit)

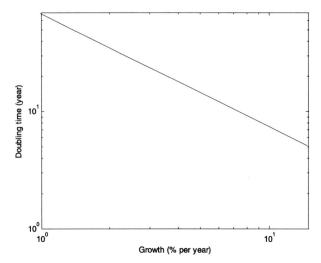

Fig. 1.5 Doubling time

Since energy cannot be created or destroyed and with the expected population increase, it is anticipated that there will be energy crises in the future, which may lead to an energy dilemma due to the finite amount of readily available fossil fuels. The population of human beings has increased in the last century by a factor of 6 but the energy consumption by a factor of 80. The worldwide average continuous power consumption today is 2 kW/person. In the USA the power consumption is on average 10 kW/person and in Europe about 5 kW/person and two billion people on earth do not consume any fossil fuels at all. The reserves of fossil fuels on earth are limited and predictions based on the continuation of the energy consumption development show that the demand will soon exceed the supply. The world's population increases at 1.3 – 2% per year so that it is expected to double within the next 60 years. According to the International Energy Agency (IEA 2000) the present population is about 6.5×10^9 and growing toward 12×10^9 in 2060. At the same time, developing countries want the same standard of living as developed countries. The world population is so large that there is an uncontrolled experiment taking place on the earth's environment. The developed countries are the major contributors to this uncontrolled experiment.

The poor, who make up half of the world's population and earn less than US$ 2 a day (UN-Habitat 2003), cannot afford adaptation mechanisms such as air-conditioning, heating, or climate-risk insurance (which is unavailable or significantly restricted in most developing countries). The poor depend on water, energy, transportation, and other public infrastructures which, when affected by climate-related disasters, are not immediately replaced (Freeman and Warner 2001).

Increases in the world population, demands on goods, technology, and the higher standard of comfort for human life all require more energy consumption and, accordingly, human beings started to ponder about additional alternative energy types.

1.3 Energy and Society

Prior to the discovery of fossil fuels, coal and water played a vital role in such a search. For instance, transportation means such as the oceangoing vessels and early trains ran on steam power, which was the combination of coal and water vapor. After the discovery of oil reserves, steam power became outmoded. Hence, it seemed in the first instance that an unparalleled energy alternative had emerged for the service of mankind. Initially, it was considered an unlimited resource but with the passage of time, limitations in this alternative were understood not only in the quantitative sense but also in the *environmental and atmospheric pollution senses.* Society is affected by climate and hence energy in one of the three major ways:

1. Economic sectors that support a settlement are affected because of changes in productive capacity or changes in market demand for the goods and services produced there (energy demand). The importance of this impact depends in part on whether the settlement is rural (which generally means that it is dependent on one or two resource-based industries with much less energy consumption) or urban, in which case there usually is a broader array of alternative resources including energy resources consumption centers.
2. Some aspects of physical infrastructure (including energy transmission and distribution systems), buildings, urban services (including transportation systems), and specific industries (such as agro-industry and construction) may be directly affected. For example, buildings and infrastructure in deltaic areas may be affected by coastal and river flooding; urban energy demand may increase or decrease as a result of changed balances in space heating and space cooling (additional energy consumption); and coastal and mountain tourism may be affected by changes in seasonal temperature and precipitation patterns and sea-level rise. Concentration of population and infrastructure in urban areas can mean higher numbers of people and a higher value of physical capital at risk, although there also are many economies of scale and proximity in ensuring a well-managed infrastructure and service provision.
3. As a result of climate change society may be affected directly through extreme weather conditions leading to changes in health status and migration. Extreme weather episodes may lead to changes in deaths, injuries, or illness. Population movements caused by climate changes may affect the size and characteristics of settlement populations, which in turn changes the demand for urban services (including energy demand). The problems are somewhat different in the largest population centers (*e.g.,* those of more than 1 million people) and mid-sized to small-sized regional centers. The former are more likely to be destinations for migrants from rural areas and smaller settlements and cross-border areas, but larger settlements generally have much greater command over national resources. Thus, smaller settlements actually may be more vulnerable. Informal settlements surrounding large and medium-size cities in the developing world remain a cause for concern because they exhibit several current health and environmental hazards that could be exacerbated by global warming and have limited command over resources.

1.4 Energy and Industry

Industry is defined as including manufacturing, transport, energy supply and demand, mining, construction, and related informal production activities. Other sectors sometimes included in industrial classifications, such as wholesale and retail trade, communications, real estate and business activities are included in the categories of services and infrastructure. An example of an industrial sector particularly sensitive to climate change is energy (Hewer 2006). After the *industrial revolution* in the mid-eighteenth century human beings started to require more energy for consumption. Hence, non-renewable energy sources in the form of coal, oil, and wood began to deplete with time. As a result, in addition to the limited extent and environmental pollution potential, these energy sources will need to be replaced by renewable alternatives.

Global net energy demand is very likely to change (Tol 2002b) as demand for air-conditioning is highly likely to increase, whereas demand for heating is highly likely to decrease. The literature is not clear on what temperature is associated with minimum global energy demand, so it is uncertain whether warming will initially increase or decrease net global demand for energy relative to some projected baseline. However, as temperatures rise, net global demand for energy will eventually rise as well (Scheinder *et al.*, 2007).

Millennium goals were set solely by indicators of changes in energy use per unit of GDP and/or by total or per capita emissions of CO_2. Tracking indicators of protected areas for biological diversity, changes in forests, and access to water all appear in the goals, but they are not linked to *climate-change impacts* or adaptation; nor are they identified as part of a country's capacity to adapt to climate change (Yohe *et al.*, 2007).

With the unprecedented increase in the population, the industrial products, and the development of technology, human beings started to search for new and alternative ways of using more and more energy without harming or, perhaps, even destroying the natural environment. This is one of the greatest unsolved problems facing mankind in the near future. There is an unending debate that the key atmospheric energy source, *solar radiation*, should be harnessed more effectively and turned directly into heat energy to meet the growing demand for cheaper power supplies.

The net return from industrial material produced in a country is the reflection of energy consumption of the society in an efficient way. Otherwise, burning fossil fuels without economic industrial return may damage any society in the long run, especially with the appearance of renewable energy resources that are expected to be more economical, and therefore, exploitable in the long run. The extensive fossil fuel reservoirs available today are decreasing at an unprecedented rate and, hence, there are future *non-sustainability* alarms on this energy source. It is, therefore, necessary to diminish their exploitation rate, even starting from today, by partial replacements, especially through the *sustainable alternatives* such as solar energy.

The fossil fuel quantities that are consumed today are so great that even minor imbalances between supply and demand cause considerable *societal disruptions*. In order to get rid of such disruptions, at least for the time being, each country

1.4 Energy and Industry

imports coal, and especially oil to cover the energy imbalances. The oil embargo by the Organization of Petroleum Exporting Countries (OPEC) in 1973, gave the first serious warning and alarm to industrialized countries that energy *self-sufficiency* is an essential part of any country concerned for its economic, social, and even cultural survival. In fact, the technological and industrial developments in the last 150 years rendered many countries to energy-dependent status.

Worldwide use of energy for several decades, especially in the industrial sectors, appeared to be increasing dramatically, but in the last decade, it has leveled off, and even dropped to a certain extent as shown in Fig. 1.6. In this graph, all forms of energy uses are represented in terms of the amount of coal that would provide the equivalent energy. Around the 1970s most of the predictions foresaw that energy demand would continue to accelerate causing expected severe energy shortages. However, just the opposite situation has developed, and today, there is a surplus of energy on the worldwide market that has resulted from economic downturn coupled with many-fold increases in the oil price during the last 20 years.

Fossil fuel reserves in the form of oil and *natural gas* are still adequate at present consumption rates for the next 50 years. However, with increasing amounts of renewable energy and discoveries of new reservoirs this span of time is expected to extend for almost a century from now onward.

Linkage systems, such as transportation and transmission for industry and settlements (*e.g.*, water, food supply, energy, information systems, and waste disposal), are important in delivering the ecosystem and other services needed to support human well-being, and can be subject to climate-related extreme events such as *floods*, landslides, fire, and severe storms.

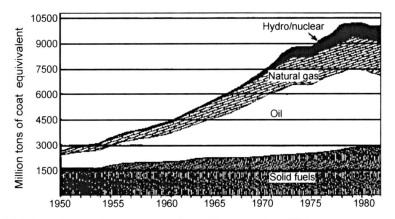

Fig. 1.6 Changes in annual energy consumption in the world (Dunn 1986)

1.5 Energy and the Economy

Continuance of *economic growth* and prosperity rely heavily on an adequate energy supply at reasonably low costs. On the other hand, energy is the main source of pollution in any country on its way to development. In general, conventional (non-renewable) energy resources are limited as compared to the present and foreseeable future energy consumptions of the world. As a whole electricity production based on fossil or nuclear fuels induces substantial social and environmental costs whereas it would appear that the use of renewable energy sources involves far less and lower costs. There are a number of different energy cost categories borne by third parties who ought to be taken into consideration in the comparison of different energy resources and technologies. Hohmeyer (1992) has given the following seven effective categories for consideration:

1. Impact on human health:
    a. Short-term impacts, such as injuries
    b. Long-term impacts, such as cancer
    c. Intergenerational impacts due to genetic damage
2. Environmental damage on:
    a. Flora, such as crops and forests
    b. Fauna, such as cattle and fish
    c. Global climate
    d. Materials
3. Long-term cost of resource depletion:
    a. Structural macro-economic impacts, such as employment effects
4. Subsidies for:
    a. Research and development
    b. Operation costs
    c. Infrastructure
    d. Evacuation in cases of accidents
5. Cost of an increased probability of wars due to:
    a. Securing energy resources (such as the Gulf War)
    b. Proliferation of nuclear weapons
6. Cost of radioactive contamination of production equipment and dwellings after major nuclear accidents
7. Psycho-social cost of:
    a. Serious illness and death
    b. Relocation of population

Adaptation strategies and implementation are strongly motivated by the cost of energy (Rosenzweig *et al.*, 2007). The nature of *adaptation* and *mitigation* decisions changes over time. For example, mitigation choices have begun with relatively easy measures such as adoption of low-cost supply and demand-side options in the energy sector (such as passive solar energy) (Levine *et al.*, 2007). Through successful investment in research and development, low-cost alternatives should become available in the energy sector, allowing for a transition to low-carbon venting pathways. Given the current composition of the energy sector, this is unlikely to happen overnight but rather through a series of decisions over time. Adaptation decisions have begun to address current *climatic risks* (*e.g.*, drought early-warning systems) and to be anticipatory or proactive (*e.g.*, land-use management). With increasing climate change, autonomous or reactive actions (*e.g.*, purchasing air-conditioning during or after a heat wave) are likely to increase. Decisions might also break trends, accelerate transitions, and mark substantive jumps from one development or technological pathway to another (Martens and Rotmans 2002; Raskin *et al.*, 2002a,b). Most studies, however, focus on technology options, costs, and competitiveness in energy markets and do not consider the implications for adaptation. For example, McDonald *et al.*(2006) use a global computed general equilibrium model and find that substituting switch grass for crude oil in the USA would reduce the GDP and increase the world price of cereals, but they do not investigate how this might affect the prospects for adaptation in the USA and for world agriculture. This limitation in scope characterizes virtually all bioenergy studies at the regional and sectorial scales, but substantial literature on adaptation-relevant impacts exists at the project level (Pal and Sharma 2001).

Other issues of particular concern include ensuring energy services, promoting agriculture and industrialization, promoting trade, and upgrading technologies. Sustainable *natural-resource management* is a key to sustained economic growth and poverty reduction. It calls for clean energy sources, and the nature and pattern of agriculture, industry, and trade should not unduly impinge on ecological health and resilience. Otherwise, the very basis of economic growth will be shattered through environmental degradation, more so as a consequence of climate change (Sachs 2005). Put another way by Swaminathan (2005), developing and employing "eco-technologies" (based on an integration of traditional and frontier technologies including biotechnologies, renewable energy, and modern management techniques) is a critical ingredient rooted in the principles of economics, gender, social equity, and employment generation with due emphasis given to climate change (Yohe *et al.*, 2007).

1.6 Energy and the Atmospheric Environment

Even though the natural circulation in the atmosphere provides scavenging effects, continuous and long-term loading of atmosphere might lead to undesirable and dangerous situations in the future. Therefore, close inspection and control should be directed toward various phenomena in the atmosphere. Among these there are more applied and detailed research needs in order to appreciate the meteorological events in the troposphere, ozone depletion in the stratosphere, pollution in the lower troposphere and trans-boundary between the troposphere and hydro-lithosphere, energy, transport and industrial pollutants generation and movement, effects of acid rain, waste water leakage into the surface, and especially ground water resources.

For success in these areas, it is necessary to have sound scientific basic research with its proper applications. The basic data for these activities can be obtained from extensive climatic, meteorological, hydrological, and hydro-geological observation network establishments with spatial and temporal monitoring of the uncontrollable variables. Ever greater cooperation is needed in detecting and predicting atmospheric changes, and assessing consequential environmental and socio-economic impacts, identifying dangerous pollution levels and greenhouse gases. New and especially renewable energy sources are required for controlling emissions of greenhouse gases. Consumption of fossil fuels in industry as well as transportation gives rise to significant atmospheric emissions. The major points in energy use are the protection of the environment, human health, and the hydrosphere. Any undesirable changes in the atmospheric conditions may endanger forests, hydrosphere ecosystems, and economic activities such as agriculture. The ozone layer within the stratosphere is being depleted by reactive chlorine and bromine from human-made *chlorofluorocarbons* (CFCs) and related substances. Unfortunately, levels of these substances in the atmosphere increase continuously signaling future dangers if necessary precautions are not taken into consideration.

It has been stated by Dunn (1986) that several problems have arisen from the increased use of energy, *e.g.*, *oil spillages* resulting from accidents during tanker transportation. Burning of various energy resources, especially fossil fuels, has caused a global-scale CO_2 rise. If the necessary precautions are not considered in the long run, this gas in the atmosphere could exceed the natural levels and may lead to climatic change. Another problem is large-scale air pollution in large cities especially during cold seasons. The use of fossil fuels in automobiles produces exhaust gases that also give rise to air pollution as well as increasing the surface ozone concentration which is dangerous for human health and the environment. Air pollution leads to acid rain that causes pollution of surface and groundwater resources which are the major water supply reservoirs for big cities.

In order to reduce all these unwanted and damaging effects, it is consciously desirable to shift toward the use of environmentally friendly and clean *renewable energy* resources, and especially, the solar energy alternatives. It seems that for the next few decades, the use of conventional energy resources such as oil, coal, and natural gas will continue, perhaps at reduced rates because of some replacement by renewable sources. It is essential to take the necessary measures and developments toward more exploitation of solar and other renewable energy alternatives by the advancement in research and technology. Efforts will also be needed in conversion and moving toward a less energy demanding way of life.

The use of energy is not without penalty, in that energy exploitation gives rise to many undesirable degradation effects in the surrounding environment and in life. It is, therefore, necessary to reduce the *environmental impacts* down to a minimum level with the optimum energy saving and management. If the energy consumption continues at the current level with the present energy sources, which are mainly of fossil types, then the prospects for the future cannot be expected to be sustainable or without negative impacts. It has been understood by all the nations since the 1970s that the energy usage and types must be changed toward more clean and environ-

1.6 Energy and the Atmospheric Environment

mentally friendly sources so as to reduce both *environmental* and *atmospheric pollutions*. *Sustainable future* development depends largely on the pollution potential of the energy sources. The criterion of sustainable development can be defined as the development that meets the needs of the present without compromising the ability of future generations to meet their own needs. *Sustainable development* within a society demands a sustainable supply of energy and an effective and efficient utilization of energy resources. In this regard, *solar energy* provides a potential alternative for future prospective development. The major areas of environmental problems have been classified by Dincer (2000) as follows:

1. Major environmental accidents
2. Water pollution
3. Maritime pollution
4. Land use and siting impact
5. Radiation and radioactivity
6. Solid waste disposal
7. Hazardous air pollution
8. Ambient air quality
9. Acid rain
10. Stratospheric ozone depletion
11. Global climate change leading to greenhouse effect

The last three items are the most widely discussed issues all over the world. The main gaseous pollutants and their impacts on the environment are presented in Table 1.2.

Unfortunately, energy is the main source of pollution in any country on its way to development. It is now well known that the sulfur dioxide (SO_2) emission from fossil fuels is the main cause of *acid rain* as a result of which more than half the forests in the Northern Europe have already been damaged. In order to decrease degradation effects on the environment and the atmosphere, technological developments have been sought since the 1973 oil crisis. It has been recently realized that

Table 1.2 Main gaseous pollutants

Gaseous pollutants	Greenhouse effect	Stratospheric ozone depletion	Acid precipitation
Carbon monoxide (CO)	+	±	
Carbon dioxide (CO_2)	+	±	
Methane (CH_4)	+	±	
Nitric oxide (NO) and nitrogen dioxide (NO_2)	±	+	+
Nitrous oxide (N_2O)	+	±	
Sulfur dioxide (SO_2)	−	+	
Chlorofluorocarbon(CFCs)	+	+	
Ozone (O_3)	+	+	

Plus and minus signs indicate proportional and inversely proportional effects whereas ± implies either effect depending on circumstances

renewable energy sources and systems can have a beneficial impact on the following essential technical, environmental, and political issues of the world. These are:

1. Major environmental problems such as acid rain, *stratospheric ozone* depletion, *greenhouse* effect, and smog
2. Environmental degradation
3. Depletion of the world's non-renewable conventional sources such as coal, oil, and natural gas
4. Increasing energy use in the developing countries
5. World population increase

In most regions, climate change would alter the probability of certain weather conditions. The only effect for which average change would be important is *sea-level rise*, under which there could be increased risk of *inundation* in coastal settlements from average (higher) sea levels. Human settlements for the most part would have to adapt to more or less frequent or intense rain conditions or more or less frequent mild winters and hot summers, although individual day weather may be well within the range of current weather variability and thus not require exceptionally costly adaptation measures. The larger, more costly impacts of climate change on human settlements would occur through increased (or decreased) probability of extreme weather events that overwhelm the designed resiliency of human systems.

Much of the urban center managements as well as the governance structures that direct and oversee them are related to reducing *environmental hazards*, including those posed by extreme weather events and other natural hazards. Most regulations and management practices related to buildings, land use, waste management, and transportation have important environmental aspects. Local capacity to limit environmental hazards or their health consequences in any settlement generally implies local capacity to adapt to climate change, unless adaptation implies particularly expensive infrastructure investment.

An increasing number of urban centers are developing more comprehensive plans to manage the environmental implications of urban development. Many techniques can contribute to better *environmental planning* and *management* including market-based tools for pollution control, demand management and waste reduction, mixed-use zoning and transport planning (with appropriate provision for pedestrians and cyclists), environmental impact assessments, capacity studies, strategic environmental plans, environmental audit procedures, and state-of-the-environment reports (Haughton 1999). Many cities have used a combination of these techniques in developing "Local Agenda 21s," which deal with a list of urban problems that could closely interact with climate change and energy consumption in the future. Examples of these problems include the following points (WRI 1996):

1. Transport and road infrastructure systems that are inappropriate to the settlement's topography (could be damaged by landslides or flooding with climate change)
2. Dwellings that are located in high-risk locations for floods, landslides, air and water pollution, or disease (vulnerable to flood or landslides; disease vectors more likely)

3. Industrial contamination of rivers, lakes, wetlands, or coastal zones (vulnerable to flooding)
4. Degradation of landscape (interaction with climate change to produce flash floods or desertification)
5. Shortage of green spaces and public recreation areas (enhanced heat island effects)
6. Lack of education, training, or effective institutional cooperation in environmental management (lack of adaptive capacity)

1.7 Energy and the Future

The world demand for energy is expected to increase steadily until 2030 according to many scenarios. Global primary energy demand is projected to increase by 1.7% per year from 2000 to 2030, reaching an annual level of 15.3×10^9 tons of oil equivalent (toe). The projected growth is, nevertheless, slower than the growth over the past 30 years, which ran at 2.1% per year. The global oil demand is expected to increase by about 1.6% per year from 75×10^6 barrels per day to 120×10^6 barrels per day. The transportation sector will take almost three quarters of this amount. Oil will remain the fuel of choice in transportation (IEA 2002).

The energy sources sought in the long term are hoped to have the following important points for a safer and more pleasant environment in the future:

1. Diversity of various alternative energy resources both conventional (non-renewable) and renewable, with a steadily increasing trend in the use of renewable resources and a steadily decreasing trend over time in the non-renewable resources usage.
2. Quantities must be abundant and sustainable in the long term.
3. Acceptable cost limits and prices compatible with strong economic growth.
4. Energy supply options must be politically reliable.
5. Friendly energy resources for the environment and climate change.
6. Renewable domestic resources that help to reduce the important energy alternatives.
7. They can support small to medium scale local industries.

The renewable energies are expected to play an active role in the future energy share because they satisfy the following prerequisites:

1. They are environmentally clean, friendly, and do not produce greenhouse gases.

2. They should have sufficient resources for larger scale utilization. For instance, the solar energy resources are almost evenly distributed all over the world with maximum possible generatable amounts increasing toward the equator.
3. The intermittent nature of solar and wind energy should be alleviated by improving the storage possibilities.

4. The cost effectiveness of the renewable sources is one of the most important issues that must be tackled in a reduction direction. However, new renewable energies are now, by and large, becoming cost competitive with conventional forms of energy.

In order to care for the future generations, energy conservation and savings are very essential. Toward this end one has to consider the following points:

1. Conservation and more efficient use of energy. Since the first energy crisis, this has been the most cost-effective mode of operation. It is much cheaper to save a barrel of oil than to discover new oil.
2. Reduce demand to zero growth rate and begin a steady-state society.
3. Redefine the size of the system and colonize the planets and space. For instance, the resources of the solar system are infinite and our galaxy contains over 100 billion stars.

Because the earth's resources are finite for the population, a change to a *sustainable society* depends primarily on renewable energy and this becomes imperative over a long time scale. The following *adaptation* and *mitigation* policies must be enhanced in every society:

1. Practice conservation and efficiency
2. Increase the use of *renewable energy*
3. Continue dependence on natural gas
4. Continue the use of coal, but include all social costs (externalities)

Regional and local polices must be the same. Efficiency can be improved in all major sectors including residential, commercial, industrial, transportation, and even the primary electrical utility industry. The most gains can be accomplished in the transportation, residential, and commercial sectors. National, state, and even local building codes will improve energy efficiency in buildings. Finally, there are a number of things that each individual can do in conservation and energy efficiency.

References

Arnell N, Tompkins E, Adger N, Delaney K (2005) Vulnerability to abrupt climate change in Europe. Technical Report 34, Tyndall Centre for Climate Change Research, Norwich

Cartalis C, Synodinou A, Proedrou M, Tsangrassoulis A, Santamouris M (2001) Modifications in energy demand in urban areas as a result of climate changes: an assessment for the southeast Mediterranean region. Energy Convers Manage 42:1647–1656

Clarke A (1988) Wind farm location and environmental impact. Network for Alternative Technology and Technology Assessments C/O EEDU, The Open University, UK

Dincer I (2000) Renewable energy and sustainable development: a crucial review. Renewable and Sustainable Energy Reviews 4:157–175

Dunn PD (1986) Renewable energies: sources, conversion and application. Peregrinus, Cambridge

Edmonds J, Reilly J (1985) Global energy: assessing the future. Oxford University Press, New York

References

Edmonds J, Reilly J (1993) A long-term global economic model of carbon dioxide release from fossil fuel use. Energy Econ 5:74

Freeman P, Warner K (2001) Vulnerability of infrastructure to climate variability: how does this affect infrastructure lending policies? Disaster Management Facility of The World Bank and the ProVention Consortium, Washington, District of Columbia

Hadley SW, Erickson DJ, Hernandez JL, Broniak CT, Blasing TJ (2006) Responses of energy use to climate change: a climate modeling study. Geophys Res Lett 33, L17703. doi:10.1029/2006GL026652

Haughton G (1999) Environmental justice and the sustainable city. J Plann Educ Res 18(3):233–243

Hewer F (2006) Climate change and energy management: a scoping study on the impacts of climate change on the UK energy industry. UK Meteorological Office, Exeter

Hill D (1974) The book of knowledge of ingenious mechanical devices. Reidel, Dordrecht

Hoel M, Kvendokk S (1996) Depletion of fossil fuels and the impact of global warming. Resour Energy Econ 18:115

Hohmeyer O (1992) The solar costs of electricity: renewable versus fossil and nuclear energy. Solar Energy 11:231–250

IEA (2000) The evolving renewable energy market. International Energy Agency, Paris: IEA/OECD. http://www.iae.org

IEA (2002) World energy outlook. International Energy Agency, Paris: IEA/OECD. http://www.iae.org

Levine M et al. (2007) Residential and commercial buildings. In: Metz B, Davidson O, Bosch P, Dave R, Meyer L (eds) Climate change 2007: mitigation of climate change. Cambridge University Press, Cambridge, UK

Martens P, Rotmans J (2002) Transitions in a Globalizing World. Swets and Zeitlinger, Lisse

McDonald S, Robinson S, Thierfelder K (2006) Impact of switching production to bioenergy crops: the switch grass example. Energy Econ 28:243–265

Nordhaus WD (1993) Reflections on the economics of climate change. J Econ Perspect 7:11

Nordhaus WD (1997) The efficient use of energy resources. Yale University Press, New Haven

Pal RC, Sharma A (2001) Afforestation for reclaiming degraded village common land: a case study. Biomass Bioenerg 21:35–42

Palz W (1994) Role of new and renewable energies in future energy systems. Int J Solar Energy 14:127–140

Raskin P, Gallopin G, Gutman P, Hammond A, Swart R (2002a) Bending the curve: toward global sustainability. A report of the Global Scenario Group. SEI Pole T. Star Series Report No. 8. Stockholm Environment Institute, Stockholm

Raskin P, Banuri R, Gallopin G, Gutman P, Hammond A, Kates R, Swart R (2002b) Great transition: the promise and lure of the times ahead. A report of the Global Scenario Group. SEI Pole Star Series Report No. 10. Stockholm Environment Institute, Boston

Rosenzweig C, Casassa G, Karoly DJ, Imeson A, Liu C, Menzel A, Rawlins S, Root TL, Seguin B, Tryjanowski P (2007) Assessment of observed changes and responses in natural and managed systems. In: Parry ML, Canziani OF, Palutikof JP, van der Linden PJ, Hanson CE (eds) Climate change 2007: impacts, adaptation and vulnerability. Cambridge University Press, Cambridge, UK, pp 79–131

Sachs JD (2005) The end of poverty: economic possibilities for our time. Penguin, New York

Schneider S, Semenov H, Patwardhan S, Burton A, Magadza I, Oppenheimer CHD, Pittock M, Rahman A, Smith JB, Suarez A, Yamin F (2007) Assessing key vulnerabilities and the risk from climate change. In: Parry ML, Canziani OF, Palutikof JP, van der Linden PJ, Hanson CE (eds) Climate change 2007: impacts, adaptation and vulnerability. Cambridge University Press, Cambridge, UK, pp 779–810

Şen Z (2005) Batmayan güneºlerimiz. Our suns without sunset (in Turkish). Bilim Serisi. Altın Burç Yayınları

Swaminathan MS (2005) Environmental education for a sustainable future. In: Singh JS, Sharma VP (eds) Glimpses of the work on environment and development in India. Angkor, New Delhi, pp 51–71
Tol RSJ (2002a) Estimates of the damage costs of climate change, part I: benchmark estimates. Environ Resour Econ 21:47–73
Tol RSJ (2002b) Estimates of the damage costs of climate change, part II: dynamic estimates. Environ Resour Econ 21:135–160
Tsur Y, Zemel A (1996) Accounting for global warming risks: resource management under event uncertainty. J Econ Dyn Control 20:1289
Tsur Y, Zemel A (1998) Pollution control in an uncertain environment. J Econ Dyn Control 22:967
UN-Habitat (2003) The challenge of slums: global report on human settlements 2003. Earthscan, London
Weyant JP (1993) Cost of reducing global carbon emission. J Econ Perspect 7:27
Wilbanks TJ, Romero Lankao P, Bao M, Berkhout F, Cairncross S, Ceron JP, Kapshe M, Muir-Wood R, Zapata-Marti R (2007) Industry, settlement and society. In: Parry ML, Canziani OF, Palutikof JP, van der Linden PJ, Hanson CE (eds) Climate change 2007: impacts, adaptation and vulnerability. Cambridge University Press, Cambridge, UK, pp 357–390
WRI (1996) World resources 1996–97. World Resources Institute, Oxford University Press, New York
Yohe GW, Lasco RD, Ahmad QK, Arnell NW, Cohen SJ, Hope C, Janetos AC, Perez RT (2007) Perspectives on climate change and sustainability. In: Parry ML, Canziani OF, Palutikof JP, van der Linden PJ, Hanson CE (eds) Climate change 2007: impacts, adaptation and vulnerability. Cambridge University Press, Cambridge, UK, pp 811–841
Zierl B, Bugmann H (2005) Global change impacts on hydrological processes in Alpine catchments. Water Resour Res 41:W02028

Chapter 2
Atmospheric Environment and Renewable Energy

2.1 General

Human beings, animals, and plants alike are dependent on some gases, nutrients, and solids that are available rather abundantly and almost freely in nature for their survival. Among these the most precious ones are the air in the atmosphere that living organisms breathe and the water that is available in the hydrosphere either in the troposphere as a vapor (humidity) or in the lithosphere as a liquid (rainfall, runoff, groundwater, seas, lakes) or a solid (glaciers, snow, ice, hail). The atmosphere has evolved over geological time and the development of life on the earth has been closely related to the composition of the atmosphere. From the geological records, it seems that about 1.5 billion years ago free oxygen first appeared in the atmosphere in appreciable quantities (Harvey 1982). The appearance of life was very dependent on the availability of oxygen but once a sufficient amount had accumulated for green plants to develop, photosynthesis was able to liberate more into the atmosphere. During all these developments solar radiation provided the sole energy source.

In general, there are six different heat and mass exchanges within the atmosphere. These exchanges play the main role in the energy distribution throughout the whole system. The major energy source is solar radiation between the atmosphere and space. This energy source initiates the movement of heat and mass energy from the oceans (seas) into the air and over the land surfaces. The next important heat energy transfer occurs between the free surface bodies (oceans, seas, rivers, reservoirs) and the atmosphere. Thus water vapor, as a result of evaporation, is carried at heights toward the land by the kinetic energy of the wind. Such a rise gives the water vapor potential energy. After condensation by cooling, the water vapor appears in the form of precipitation and falls at high elevations forming the surface runoff which due to gravity flows toward the seas. During its travel toward the earth's surface, a raindrop loses its potential energy while its kinetic energy increases. Water vapor is the inter-mediator in such a dynamic system. Finally, the water is returned to the seas via streams and rivers, because gravity ultimately takes over the movement of masses. The natural energy cycle appears as an integral part of the hydrological cycle (Şen 1995). During this cycle, no extra energy is produced within the atmosphere. Such movements result from the fine balance that has existed for so long between the output of radiation from the sun and the overall effects of the earth's gravitation.

Groundwater and surface water bodies become acidified due to trans-boundary air pollution causing harm to human health, and tree and forest loss. Unfortunately, there is not enough data for assessment of these dangers in the developing countries. One of the greatest and most famous scientist all over the world Ibn Sina (Avicenna, 958–1037) recommended some 1000 years ago seven points for a human being to sustain a healthy life in this world (Şen 2005). These are:

1. Spiritual healthiness
2. Choice of food and drinking water quality
3. Getting rid of extra weight to feel fit
4. Healthiness of the body
5. Comfortable dressing
6. Cleanliness of the inhaled air
7. Healthiness in thinking and pondering

Two of these points, namely, choice of water and air clarity will constitute the main topic of this chapter related to renewable energy sources in general but to solar energy in particular.

2.2 Weather, Climate, and Climate Change

Weather describes the short-term (*i. e.*, hourly and daily) state of the atmosphere. It is not the same as *climate*, which is the long-term average weather of a region including typical weather patterns, the frequency and intensity of storms, cold spells, and heat waves. However, *climate change* refers to changes in long-term trends in the average climate, such as changes in average temperatures. In Intergovernmental Panel on Climate Change (IPCC) terms, climate change refers to any change in climate over time, whether due to natural variability or as a result of human activity. Climate *variability* refers to changes in patterns, such as precipitation patterns, in the weather and climate. Finally, the greenhouse effect (*global warming*) is a progressive and gradual rise of the earth's average surface temperature thought to be caused in part by increased concentrations of CFCs in the atmosphere.

In the past, there have been claims that all weather and climate changes are caused by variations in the solar irradiance, countered at other times by the assertion that these variations are irrelevant as far as the lower atmosphere is concerned (Trenberth *et al.*, 2007). The existence of the atmosphere gives rise to many atmospheric and meteorological events. Greenhouse gases are relatively transparent to visible light and relatively opaque to infrared radiation. They let sunlight enter the atmosphere and, at the same time, keep radiated heat from escaping into space. Among the major greenhouse gases are carbon dioxide (CO_2), methane (CH_4), and nitrous oxide (N_2O), which contribute to global warming (climate change) effects in the atmosphere. Atmospheric composition has changed significantly since pre-industrial times and the CO_2 concentration has risen from 280 parts per million (ppm) to around 370 ppm today, which corresponds to about a 0.4% increase per year. On the other hand, CH_4 concentration was about 700 parts per billion (ppb) but has reached 1700 ppb today, and N_2O has increased from 270 ppb to over 310 ppb. Halocarbon does not exist naturally in the atmosphere, but since the 1950s it has accumulated in appreciable amounts causing noticeable greenhouse effects. These concentration increases in the atmosphere since the 1800s are due almost entirely to human activities.

The amount of the solar radiation incident on the earth is partially reflected again into the earth's atmosphere and then onward into the space. The reflected amount is referred to as the planetary *albedo*, which is the ratio of the reflected (scattered) solar radiation to the incident solar radiation, measured above the atmosphere. The amount of solar radiation absorbed by the atmospheric system plays the dominant role in the generation of meteorological events within the lower atmosphere (troposphere) and for the assessment of these events the accurate determination of *planetary albedo* is very important. The absorbed solar energy has maximum values of 300 W/m^2 in low latitudes. On the basis of different studies, today the average global albedo is at about 30% with maximum change of satellite measurement at ±2%, which is due to both seasonal and inter-annual time scales. Furthermore, the maximum (minimum) values appear in January (July). The annual variations are as a result of different cloud and surface distributions in the two hemispheres. For instance, comparatively more extensive snow surfaces are present in the northern European and Asian land masses in addition to a more dynamic seasonal cycle of clouds in these areas than the southern polar region. Topography is the expression of the earth's surface appearance, height, and surface features. It plays an effective role both in the generation of meteorological events and solar radiation distribution. Although the *surface albedo* is different than the planetary albedo, it makes an important contribution to the planetary albedo. The cloud distribution is the major dominant influence on the earth surface incident solar energy. Since the albedo is a dominant factor in different meteorological and atmospheric events, its influence on the availability of solar radiation has an unquestionable significance. The calculation of solar energy potential at a location is directly related to albedo-affected events and the characteristics of surface features become important (Chap. 3). In general, the albedo and hence the solar radiation energy potential at any location is dependent on the following topographical and morphological points:

1. The type of surface
2. The solar elevation and the geometry of the surface (horizontal or slope) relative to the sun
3. The spectral distribution of the solar radiation and the spectral reflection

Table 2.1 indicates different surface albedo values with the least value being for a calm sea surface at 2%, and the maximum for a fresh snow surface reaching up to 80%. In general, forests and wet surfaces have low values and snow-covered

Table 2.1 Albedo values

Surface	Albedo (%)
New snow	85
Old snow	75
Clayey desert	29–31
Green grass	8–27
Pine forest	6–19
Calm sea surface	2–4
Granite	12–18
Water (depending on angle of incidence)	2–78
High-level cloud	21
Middle-level cloud (between 3 and 6 km)	48
Low-level cloud sheets	69
Cumulus clouds	70

surfaces and deserts have high albedo values. On the other hand, the surface albedo is also a function of the spectral reflectivity of the surface.

The planetary radiation is dominated by emission from the lower troposphere. It shows a decrease with latitude and such a decrease is at a slower rate than the decrease in the absorbed solar radiation energy. At latitudes less than 30° the planetary albedo is relatively constant at 25% and, consequently, there are large amounts of solar radiation for solar energy activities and benefits in these regions of the earth. However, the solar absorption exceeds the planetary emission between 40° N and 40° S latitudes, and therefore, there is a net excess in low latitudes and a net deficit in high latitudes. Consequently, such an imbalance in the solar radiation energy implies heat transfer from low to high latitudes by the circulations within the atmosphere. Accordingly, the solar energy facilities decrease steadily from the equatorial region toward the polar regions. It is possible to state that the natural atmospheric circulations at planetary scales are due to solar energy input into the planetary atmosphere. In order to appreciate the heat transfer by the atmosphere, the difference between the absorbed and emitted planetary solar radiation amounts can be integrated from one pole to other, which gives rise to *radiation change* as in Fig. 2.1. It can be noted that the maximum transfer of heat occurs between 30° and 40° of latitude and it is equal to 4×10^{15} W.

The *regional change* of net solar *radiation* budget is shown in Fig. 2.2, which indicates substantial seasonal variation.

Increased cloudiness can reduce solar energy production. For many reasons, clouds are critical ingredients of climate and affect the availability of many renewable energy resources at a location (Monteith 1962). About half of the earth is covered by clouds at any instant. The clouds are highly dynamic in relation to atmospheric circulation. Especially, the irradiative properties of clouds make them a key component of the earth's *energy budget* and hence solar energy.

2.2 Weather, Climate, and Climate Change

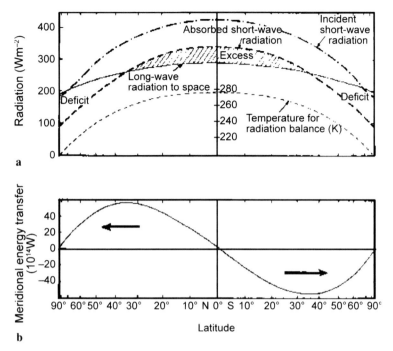

Fig. 2.1 a,b. Zonal solar radiation changes

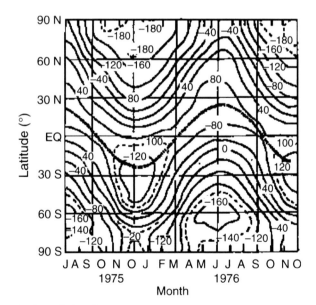

Fig. 2.2 Seasonal solar radiation changes

2.3 Atmosphere and Its Natural Composition

Foreign materials that man releases into the atmosphere at least temporarily and locally change its composition. The most significant man-made atmospheric additions (carbon monoxide, sulfur oxides, hydrocarbons, liquids such as water vapor, and solid particles) are gases and aerosol particles that are toxic to animal and plant life when concentrated by local weather conditions, such as inversion layer development, orographic boundaries, and low pressure areas. The principal pollution sources of toxic materials are automotive exhausts and sulfur-rich coal and petroleum burned for power and heating. Fortunately, most toxic pollutants are rather quickly removed from the atmosphere by natural weather processes depending on the meteorological conditions, which do not have long-term effects as explained in the previous section.

In a pollutant intact atmosphere naturally available gases namely, nitrogen, oxygen, and carbon dioxide are replenished through cycles lasting many years due to the natural phenomena that take place between various spheres. Figure 2.3 shows the interaction between the atmosphere, biosphere, lithosphere, and hydrosphere for the *nitrogen cycle* that is the main constituent in the atmosphere and it completes its renewal process about once every 100 million years. Nitrogen is the dominant element in the lower atmosphere (about 78%) but it is among the rarer elements both in the hydrosphere and the lithosphere. It is a major constituent not only of the atmosphere, but also of the animals and plants of the living world where it is a principal element in proteins, the basic structural compounds of all living organisms. Certain microscopic bacteria convert the tremendous nitrogen supply of the atmosphere into water-soluble nitrate atom groups that can then be used by plants and animals for protein manufacturing. The nitrogen re-enters the atmosphere as dead animals and plants are decomposed by other nitrogen-releasing bacteria.

The second major constituent of the lower atmosphere is oxygen (about 21%), which is the most abundant element in the hydrosphere and lithosphere. Most of the uncombined gaseous oxygen of the atmosphere is in neither the hydrosphere nor the lithosphere but as a result of *photosynthesis* by green plants. In the photosynthesis process sunlight breaks down water into hydrogen and oxygen. The free oxygen is utilized by animals as an energy source being ultimately released into the atmosphere combined with carbon as CO_2, which is taken up by plants to begin the cycle again, as shown in Fig. 2.4. Such a cycle recycles all the *oxygen* available in the atmosphere in only 3000 years. Thus the free oxygen like nitrogen is closely interrelated with the life processes of organisms.

Although CO_2 is one of the minor constituents of the lower atmosphere, it plays a fundamental role in the atmospheric heat balance, like *ozone* within the stratosphere, and is a major controlling factor in the earth's patterns of weather and climate. The CO_2 cycle is shown schematically in Fig. 2.5.

Green plants directly use atmospheric CO_2 to synthesize more complex carbon compounds which, in turn, are the basic food for animals and non-green plants. The carbon is ultimately returned into the atmosphere as a waste product of animal and plant respiration or decomposition just as free oxygen is contributed by green plant

2.3 Atmosphere and Its Natural Composition

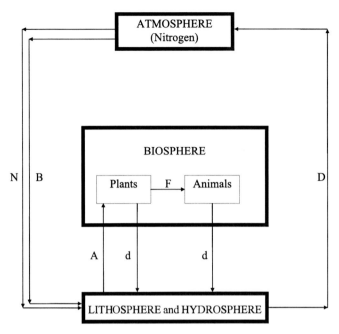

Fig. 2.3 Natural nitrogen cycle. *N* nitrogen compounds produced by atmospheric electrical storms, *B* bacteria, *A* assimilation, *d* death, *F* feeding, *D* decomposition of organic matter

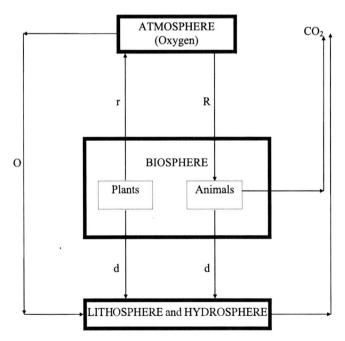

Fig. 2.4 Natural oxygen cycle. *d* death, *O* oxidation of dead organic matter, *r* released by photosynthesis, *R* respiration

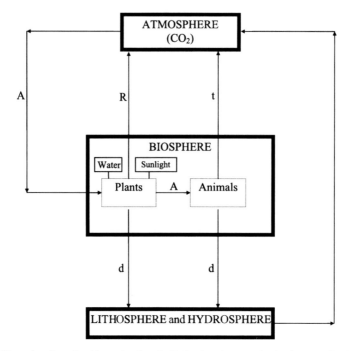

Fig. 2.5 Natural carbon dioxide cycle. A assimilation by photosynthesis, d death, O oxidation of dead organic matter, r released by photosynthesis, R respiration

photosynthesis. It takes only 35 years for the relatively small quantity of CO_2 in the atmosphere to pass once through this cycle.

2.4 Anthropogenic Composition of the Atmosphere

Climate change due to the use of CFCs is a major cause of imbalance and natural absorption of CO_2 is another example of possible social costs from energy use, which are handed over to future generations by today's energy consumers. Again the major source of *climate change* is the poor combustion of fossil fuels.

The atmosphere functions like a blanket, keeping the earth's heat from radiating into space. It lets solar insolation in, but prevents most of the ground infrared radiation from going out. The greenhouse gases are CO_2, N_2O, NH_4, water vapor, and other trace gases such as methane. A large atmosphere with a high concentration of CO_2 can drastically change the energy balance. The greenhouse effect is amply demonstrated on a sunny day by any car interior with the windows closed. The incident light passes through the windows and is absorbed by the material inside, which then radiates (infrared) at the corresponding temperature. The windows are opaque to infrared radiation and the interior heats up until there is again an energy balance.

2.4 Anthropogenic Composition of the Atmosphere

There is an increase in CO_2 in the atmosphere due to the increased use of fossil fuels and many scientists say that this results in global warming. The same thing is now said about global warming as was said about the ozone problem. It is not quite possible to reduce the production of CO_2, because of economics and the science for CO_2 and global warming is not completely certain.

2.4.1 Carbon Dioxide (CO_2)

The consumption of the fossil fuels is responsible for the increase of the CO_2 in the atmosphere by approximately 3×10^{12} kg/year (IPCC 2007). CO_2 is a greenhouse gas and causes an increase in the average temperature on earth. The major problem is the fact that a large amount, approximately 98% of CO_2 on earth, is dissolved in the water of the oceans (7.5×10^{14} kg in the atmosphere, 4.1×10^{16} kg in the ocean). The solubility of CO_2 decreases with the increasing temperature of water by approximately 3%/degree Kelvin. If the average temperature of the oceans increases the CO_2 solubility equilibrium between the atmosphere and the oceans shifts toward the atmosphere and then leads to an additional increase in the greenhouse gas in the atmosphere. The world eco-system is suffering from air pollution and global warming. The issue is a central problem now for every evolving technology to be accepted by the global community. It is therefore necessary to develop new and eco-friendly technologies. The global system is being disturbed such that it is no longer tolerant of further dirty technologies.

As a result of burning coal and oil as fuel, the level of CO_2 has risen significantly in the last 100 years. It is estimated that CO_2 accounts for about 60% of the *anthropogenic* (or human-caused) greenhouse change, known as the *enhanced greenhouse effect*. If carbon fuels are of biological origin, then sometime in the earth's distant past there must have been far more CO_2 in the atmosphere than there is today. There are two naturally different sources for this gas, as emissions from animal life and decaying plant matter, *etc.* (Fig. 2.5), which constitute about 95% of the CO_2, and the rest comes from human activity (anthropogenic) sources, including the burning of carbon-based (especially fossil) fuels. It is known that although the anthropogenic share is a comparatively small portion of the total, it contributes in an accumulative manner over time.

Since CO_2 is one of the large gas molecules that traps long-wave radiation to warm the lower atmosphere by the so-called *greenhouse effect*, atmospheric scientists and meteorologists alike suggested that increase in the CO_2 might be causing a general warming of the earth's climate (IPCC 2007). Worries about the effect of CO_2 on the climate have given rise to further detailed studies and investigations to focus attention on the complex interactions between man's activities and the atmosphere that surrounds them and thus may prevent still more serious problems from arising in the future.

Although pollutants may originate from natural or man-made activities, the term *pollution* is often restricted to considerations of *air quality* as modified by human actions, particularly when pollutants are emitted from industrial, urban, commercial, and nuclear sites at rates in excess of the natural diluting and self-purifying processes currently prevailing in the lower atmosphere. Air pollution seems to be a local problem with three distinctive geographic factors:

1. All the wealth of human beings is defined by the distribution of housing, industry, commercial centers, and motor vehicle transportation between various centers. Such a system forms the major source of man-made pollution.
2. A natural phenomenon in the atmosphere, which controls the local and temporal climatic weather variations as a result of which the pollutants introduced into the atmosphere are either scattered in various directions or carried away by air movements in the form of winds.
3. The interaction between the pollution emissions and the atmosphere may well be modified by local relief factors, especially when pollution is trapped by relatively stagnant air within a valley.

Over millions of years, much of the CO_2 was removed by sea and land flora (plants). Most was returned to the air when the plant material decayed, but some of the carbon was locked up in the form of wood, peat, coal, petroleum, and natural gas (Fig. 2.5). Now that these fuels are burnt, the carbon is being released into the atmosphere once again. CO_2 is less soluble in warmer water than in cold, and as ocean surface layers warm, CO_2 could be driven out of solution and into the atmosphere, thus exacerbating the problem. There are benefits to increased CO_2 concentrations as well as potential problems. Plants need CO_2 and the optimal concentration for most plants it is estimated to be between 800 and 1200 ppm. Some plants do best at even higher concentrations and for instance the optimal range for rice is 1500 to 2000 ppm. As the atmosphere becomes richer in CO_2, crops and other plants will grow more quickly and profusely. A doubling of CO_2 concentrations can be expected to increase global crop yields by 30% or more. Higher levels of CO_2 increase the efficiency of *photosynthesis*, and raise plants' water-use efficiency by closing the pores (stomata) through which they lose moisture. The CO_2 effect is twice that for plants that receive inadequate water than for well-watered plants. In addition, higher CO_2 levels cause plants to increase their fine root mass, which improves their ability to take in water from the soil (Bradley and Fulmer 2004).

2.4.2 Methane (CH_4)

While it is 25 times more powerful a warming agent than CO_2, *methane* has a much shorter life span and its atmospheric concentration is only about 17 ppm. Concentrations have more than doubled since 1850, though for reasons that are still unclear, they have leveled off since the 1980s. Human activity accounts for about 60% of CH_4 emissions, while the rest comes from natural sources such as wetlands. Human sources include leakage from pipelines, evaporation from petroleum recovery and refining operations, rice fields, coal mines, sanitary landfills, and waste from domestic animals. About 20% of the total human greenhouse impact is due to CH_4.

2.4.3 Nitrous Oxide (N_2O)

Its warming potential is some 300 times that of CO_2. It has an atmospheric concentration of about 0.32 ppm, up from 0.28 ppm in 1850. In the United States, 70% of man-made *nitrous oxide* emissions come from the use of nitrogen-containing agricultural fertilizers and automobile exhaust. Globally, fertilizers alone account for 70% of all emissions. The Environmental Protection Agency (EPA) has calculated that production of nitrous oxide from vehicles rose by nearly 50% between 1990 and 1996 as older cars without converters were replaced with newer, converter-equipped models. Critics argue that the EPA's numbers are greatly exaggerated. In addition, they point out that converters reduce emissions of another greenhouse gas, ozone, as well as carbon monoxide and NOx (which leads to smog).

2.4.4 Chlorofluorocarbons (CFCs)

These are powerful *global warming* gases that do not exist in nature but are produced by humans. In the upper atmosphere or stratosphere, CFCs are broken down by sunlight. The chlorine that is released by this decomposition acts as a catalyst to break naturally occurring ozone (O_3) molecules into oxygen (O_2) molecules. Ozone helps block the sun's ultraviolet radiation, which can cause skin cancer after long-term exposure. Worldwide CFC emissions have been steadily dropping, and it is expected that *ozone* depletion (the ozone hole), which reached its peak in the last decade, will drop to zero later this century. The fact that CFCs are chemically inert (that is, they do not react with other chemicals) makes them very useful in a wide variety of applications, but it also means that they last for a very long time in the atmosphere, about 50,000 years.

These gases affect the climate in different ways depending upon their location in the atmosphere. At lower altitudes, they trap heat like other greenhouse gases and have a much stronger warming effect than CO_2. In fact, some can trap as much as 10,000 times more heat per molecule than CO_2. While CO_2 is measured in atmospheric concentrations of parts per million, CFCs are measured in parts per trillion. Despite their low concentrations, it is believed that these gases account for about 15% of the human greenhouse change (Bradley and Fulmer 2004).

2.4.5 Water Vapor (H_2O)

Almost 70% of the earth's surface is covered by water bodies which are referred to collectively as the *hydrosphere*. Although there is not extensive human activity in the hydrosphere itself, the intensive activities on the land threaten the biological richness of oceans and especially the coastal areas where about 60% of the world's population live. Although legislative measures are taken, their application cannot be achieved due to lack of reliable data, planning, management, international coordination, and technology transfer, and inadequate funds. The hydrosphere is polluted by sewage, agricultural chemicals, litter, plastics, radioactive substances, fertilizers, oil spills, and hydrocarbons. Land-borne pollution gets into the major hydrosphere through rivers and atmosphere. The hydrosphere is vulnerable to climate and atmospheric change including ozone depletion.

Water vapor accounts for about 94% of the natural greenhouse effect. The impact of atmospheric water vapor on the climate is complex and not well understood. It can both warm and cool the atmosphere. When water evaporates, it cools the surface from which it evaporates. In addition, heavy clouds block sunlight and reflect it back into space. On the other hand, thin cirrus clouds may tend to let solar energy in while keeping radiated energy from escaping into space. Also, moist air retains more heat than does dry air, so a humid atmosphere should be warmer than a dry one. On balance, it is believed that water vapor has a net warming effect.

The main concern about increased concentrations of atmospheric water vapor is the possibility of a strong *positive feedback* effect. As the climate warms more water evaporates thus increasing the amount of water vapor in the air. The increased concentration will, in turn, further warm the climate leading to a still higher level of water vapor in the atmosphere. This iterative cycle could go out of control, leading to damaging or even catastrophic temperature increases. However, there may be natural mechanisms to keep the climate in balance. The water vapor originates as a result of evapo-transpiration and moves between various spheres as shown in Fig. 2.6.

The transition of water from one environment to another and its transfer among them does occur naturally in the universe continuously with time. In nature, water movement depends on these transitions as well as transfers. The driving forces of such movements are the sun's radiation (solar energy) and earth's gravity. The col-

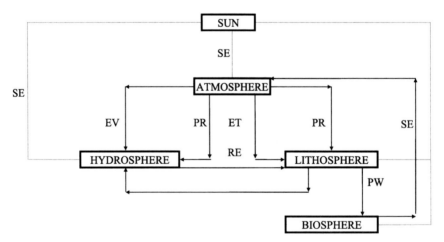

Fig. 2.6 Water environments. *EV* evaporation, *ET* evapo-transpiration, *PR* precipitation, *PW* plant water use, *RE* recharge, *SE* solar energy, *TR* transpiration

lection of these endless movement routes is referred to as the *hydrological cycle*, which is the only source that brings water as precipitation from the atmosphere to the lithosphere that includes the groundwater reservoirs. It is a gift to human beings that all of the inter-spherical connections occur with no cost at all, but at places in a very unpredictable manner as well as in rare amounts (Şen 1995).

In lakes and reservoirs, climate change effects are mainly due to water temperature variations, which result directly from climate change or indirectly through an increase in thermal pollution as a result of higher demands for cooling water in the energy sector. There is also a link between measures to adapt water resources and policies to reduce energy use. Some adaptation options, such as desalination or measures which involve pumping large volumes of water, use large amounts of energy and may be inconsistent with mitigation policy (Kundzewicz *et al.*, 2007).

Hydrology and energy regimes are two key factors that influence the coastal zonation of the plant species which typically grade inland from salt, to brackish, to freshwater species. Climate change will likely have its most pronounced effects on brackish and freshwater marshes in the coastal zone through alteration of hydrological regimes (Baldwin *et al.*, 2001; Burkett and Kusler 2000; Sun *et al.*, 2002), specifically, the nature and variability of the hydro-period and the number and severity of extreme events.

2.4.6 Aerosols

Aerosols are important to the climate system for many reasons because they have a direct effect on heating and photolysis rates in the atmosphere by scattering and absorbing radiation. They also influence the climate system indirectly by modulating cloud drop size, cloud lifetime, and precipitation in addition to many other processes such as the fuzzy effect involving subtle modulations of the dynamic and physical processes of the atmosphere. Aerosols act also on other components of the climate system by reducing the energy reaching the surface, and by transporting nutrients from one place to another. There are well-documented changes in aerosol distribution due to anthropogenic activities during the last hundred years and some changes are still expected. Solid particles in the atmosphere, namely, aerosols, constitute a significant percentage of the air pollutants. They are also referred to as the *particulate matter* that originates from other sources including ocean salt, volcanic ash, products of wind erosion, roadway dust, products of forest fires, and plant pollen and seeds. Although particulate matter comprises about 10 – 15% of the total mass of man-made air pollutants, their potential hazards are much greater. They present a health hazard to the lungs, enhance air pollution chemical reactions in the atmosphere, reduce visibility, increase the precipitation possibility in addition to fog and cloud formation, and reduce the solar radiation, leading to concomitant changes in the atmospheric environment temperature affecting the biological rates of plant growth and soil materials extensively. A detailed account of particulate matter is presented by Wark and Warner (1981).

Concentrations of air pollutants in general and fine particulate matter in particular may change in response to climate change because their formation depends, in part, on temperature and humidity. Air-pollution concentrations are the result of interactions between variations in the physical and dynamic properties of the atmosphere on time scales from hours to days, atmospheric circulation features, wind, topography, and energy use. Some air pollutants demonstrate weather-related seasonal cycles (Confalonieri *et al.*, 2007).

2.5 Energy Dynamics in the Atmosphere

With the unprecedented increase in the population and industrial products and the development of technology, human beings search for ways of using more and more energy without harming or, perhaps, even destroying the natural environment. This is one of the greatest unsolved problems facing humankind in the near future. There is an unending debate that key atmospheric energy sources such as *solar* and *wind power* should be harnessed more effectively and turned directly into heat energy to meet the growing demand for cheaper power supplies.

All types of energy can be traced back to either atmospheric activities in the past or related to the present and future activities within the atmosphere. The *renewable* or *primary energy* sources are regarded as the ones that are related to present atmospheric movements, however, secondary energy sources are *non-renewable* and have been deposited in the depths of the earth typically in the form of oil and coal. They are also referred to as the *fossil fuels*. So, burning the fossil fuels, the stored energy of past atmospheric activities, is added to the present energy demand. Consequently, their burning leads to the altering of the weather in the short term and climate in the long term in an unusual manner.

The renewable energy sources are primary energy alternatives that are part of the everyday weather elements such as sunshine and the wind. External sources of energy to the atmosphere are the sun's radiation (insolation) and the sun and moon's gravitation in the form of tides. Additionally, the internal sources are the earth's heat through conduction and earth's gravitation and rotation. These internal and external sources are constant energy supplies to the atmosphere. Apart from balancing each other they both contain thermal and mechanical forms depending on heat and mass, respectively. The *sun's radiation* (insolation) is the main source of heat energy and earth's motion and gravitation exert most influence on the masses. The atmosphere is fed by a continuous flux of radiation from the sun and gravitation remains a constant force internally.

Briefly, *radiation* is the transfer of energy through matter or space by electric or magnetic fields suitably called electromagnetic waves. High-energy waves are emitted from the tiniest particles in the nucleus of an atom, whereas low energy is associated with larger whole atoms and molecules. The highest energy waves are known as radioactivity since they are generated by the splitting (*fission*) or joining (*fusion*) of particles, and low energy waves result from vibration and collision of molecules. The sun can be regarded as a huge furnace in which hydrogen atoms fuse into helium at immensely high temperatures (Chapter 3). The solar radiation is partially absorbed by matter of increasing size, first by exciting electrons as in ionization and then by simulating molecular activity at lower energy levels. The latter is sensed as heat. Hence, radiation is continuously degraded or dissipated from tiny nuclear particles to bigger molecules of matter.

2.6 Renewable Energy Alternatives and Climate Change

Renewable energy sources are expected to become economically competitive as their costs already have fallen significantly compared with conventional energy sources in the medium term, especially if the massive subsidies to nuclear and fossil forms of energy are phased out. Finally, new renewable energy sources offer huge benefits to developing countries, especially in the provision of energy services to the people who currently lack them. Up to now, the renewable sources have been completely discriminated against for economic reasons. However, the trend in recent years favors the renewable sources in many cases over conventional sources.

The advantages of renewable energy are that they are sustainable (non-depletable), ubiquitous (found everywhere across the world in contrast to fossil fuels and minerals), and essentially clean and environmentally friendly. The disadvantages of renewable energy are its variability, low density, and generally higher initial cost. For different forms of renewable energy, other disadvantages or perceived problems are pollution, odor from biomass, avian with wind plants, and brine from geothermal.

In contrast, fossil fuels are stored solar energy from past geological ages. Even though the quantities of oil, natural gas, and coal are large, they are finite and for the long term of hundreds of years they are not sustainable. The world energy demand depends, mainly, on fossil fuels with respective shares of petroleum, coal, and natural gas at 38%, 30%, and 20%, respectively. The remaining 12% is filled by the non-conventional energy alternatives of hydropower (7%) and nuclear energy (5%). It is expected that the world oil and natural gas reserves will last for several decades, but the coal reserves will sustain the energy requirements for a few centuries. This means that the fossil fuel amount is currently limited and even though new reserves might be found in the future, they will still remain limited and the rate of energy demand increase in the world will require exploitation of other renewable alternatives at ever increasing rates. The desire to use *renewable energy* sources is not only due to their availability in many parts of the world, but also, more empathetically, as a result of the fossil fuel damage to environmental and atmospheric cleanness issues. The search for new alternative energy systems has increased greatly in the last few decades for the following reasons:

1. The extra demand on energy within the next five decades will continue to increase in such a manner that the use of fossil fuels will not be sufficient, and therefore, the deficit in the energy supply will be covered by additional energy production and discoveries.

2. Fossil fuels are not available in every country because they are unevenly distributed over the world, but renewable energies, and especially *solar radiation*, are more evenly distributed and, consequently, each country will do its best to research and develop their own national energy harvest.
3. Fossil fuel combustion leads to some undesirable effects such as *atmospheric pollution* because of the CO_2 emissions and *environmental problems* including air pollution, acid rain, greenhouse effect, climate changes, oil spills, *etc.* It is understood by now that even with refined precautions and technology, these undesirable effects can never be avoided completely but can be minimized. One way of such minimization is to substitute at least a significant part of the fossil fuel usage by *solar energy*.

In fact, the worldwide environmental problems resulting from the use of fossil fuels are the most compelling reasons for the present vigorous search for future alternative energy options that are renewable and *environmentally friendly*. The renewable sources have also some disadvantages, such as being available intermittently as in the case of solar and wind sources or fixed to certain locations including hydropower, geothermal, and biomass alternatives. Another shortcoming, for the time being, is their transportation directly as a fuel. These shortcomings point to the need for intermediary energy systems to form the link between their production site and the consumer location, as already mentioned above. If, for example, heat and electricity from solar power plants are to be made available at all times to meet the demand profile for useful energy, then an energy carrier is necessary with storage capabilities over long periods of time for use when solar radiation is not available (Veziroğlu 1995).

The use of conventional energy resources will not be able to offset the energy demand in the next decades but steady increase will continue with undesirable environmental consequences. However, newly emerging renewable alternative energy resources are expected to take an increasing role in the energy scenarios of the future *energy consumption*s.

2.6.1 Solar Energy

Almost all the renewable energy sources originate entirely from the sun. The sun's rays that reach the outer atmosphere are subjected to absorption, reflection, and transmission processes through the atmosphere before reaching the earth's surface (Chap. 3). On the other hand, depending on the earth's surface topography, as explained by Neuwirth (1980), the solar radiation shows different appearances.

The emergence of interest in solar energy utilization has taken place since 1970, principally due to the then rising cost of energy from conventional sources. *Solar radiation* is the world's most abundant and permanent energy source. The amount of solar energy received by the surface of the earth per minute is greater than the energy utilization by the entire population in one year. For the time being, solar energy, being available everywhere, is attractive for stand-alone systems particularly in the rural parts of developing nations. Occurrences of solar energy dynamically all over the world in the forms of wind, wave, and hydropower through the hydrological cycle provide abilities to ponder about their utilization, if possible instantly or in the form of reserves by various conversion facilities and technologies. It is also possible that in the very long term, human beings might search for the conversion of ocean currents and temperature differences into appreciable quantities of energy so that the very end product of solar radiation on the earth will be useful for sustainable development.

The design of many technical apparatuses such as coolers, heaters, and solar energy electricity generators in the form of photovoltaic cells, requires *terrestrial irradiation* data at the study area. Scientific and technological studies in the last three decades tried to convert the continuity of solar energy into sustainability for the human comfort. Accurate estimations of global solar radiation need meteorological, geographic, and astronomical data (Chap. 3), and especially, many estimation models are based on the easily measurable sunshine duration at a set of meteorology stations (Chaps. 4, 5, and 6).

Solar energy is referred to as renewable and/or *sustainable energy* because it will be available as long as the sun continues to shine. Estimates for the life of the main stage of the sun are another 4 – 5 billion years. The energy from the sunshine, electromagnetic radiation, is referred to as insolation.

Wind energy is derived from the uneven heating of the earth's surface due to more heat input at the equator with the accompanying transfer of water by evaporation and rain. In this sense, rivers and dams for hydro-electric energy are stored solar energy. The third major aspect of solar energy is its conversion into biomass by *photosynthesis*. Animal products such as whale oil and biogas from manure are derived from solar energy.

2.6.2 Wind Energy

It is one of the most significant and rapidly developing renewable energy sources all over the world. Recent technological developments, fossil fuel usage, environmental effects, and the continuous increase in the conventional energy resources have reduced *wind energy* costs to economically attractive levels, and consequently, wind energy farms are being considered as an alternative energy source in many enterprises.

Although the amount of wind energy is economically insignificant for the time being in many parts of the world, mankind has taken advantage of its utilization for many centuries whenever human beings found the chance to provide power for various tasks. Among these early utilizations are the hauling of water from a lower to a higher elevation (Chap. 1), grinding grains in mills by water and other mechanical power applications. It is still possible to see in some parts of the world these types of marginal benefits from wind speed. All previous activities have been technological and the scientific investigation of wind power formulations and accordingly development of modern technology appeared after the turn of the twentieth century (Şahin 2004). In recent decades the significance of wind energy has originated from its friendly behavior to the environment so far as *air pollution* is concerned, although there is, to some extent, noise and appearance pollution from the modern wind farms. Due to its cleanness, wind power is sought wherever possible for conversion to electricity with the hope that the air pollution as a result of fossil fuel burning will be reduced (Clark 1988). In some parts of the USA, up to 20% of electrical power is generated from wind energy. After the economic crisis of 1973 its importance increased due to economic factors and today there are wind farms in many western European countries (Anderson 1992; EWEA 1991; Troen and Peterson 1989).

Although the technology in converter-turbines for harnessing the wind energy is advancing rapidly, there is a need to assess its accurate behavior with scientific approaches and calculations. An effective formulation is given by Şen (2003) on a physical basis to understand, refine, and predict the variations in wind energy calculations.

Wind power is now a reliable and established technology which is able to produce electricity at costs competitive with coal and nuclear power. There will be a small increase in the annual wind energy resource over the Atlantic and northern Europe, with more substantial increases during the winters by 2071 to 2100 (Pryor *et al.*, 2005).

2.6.3 Hydropower Energy

Hydropower is an already established technological way of renewable energy generation. In the industrial and surface water rich countries, the full-scale development of *hydroelectric energy* generation by turbines at large-scale *dams* is already exploited to the full limit, and consequently, smaller hydro systems are of interest in order to gain access to the marginal resources. The world's total annual rainfall is, on average, 108.4×10^{12} tons/year of which 12×10^{12} tons recharges the groundwater resources in the aquifers, 25.13×10^{12} tons appears as surface runoff, and 71.27×10^{12} tons evaporates into atmosphere. If the above rainfall amount falls from a height of 1000 m above the earth surface, then kinetic energy of 1.062×10^{15} kJ is imparted to the earth every year. Some of this huge amount of energy is stored in dams, which confine the potential energy so that it can be utilized to generate hydroelectric power.

Wilbanks *et al.* (2007) stated that hydropower generation is likely to be impacted because it is sensitive to the amount, timing, and geographical pattern of precipitation as well as temperature (rain or snow, timing of melting). Reduced stream flows are expected to jeopardize hydropower production in some areas, whereas greater stream flows, depending on their timing, might be beneficial (Casola *et al.*, 2005; Voisin *et al.*, 2006). According to Breslow and Sailor (2002), climate variability and long-term climate change should be considered in siting wind power facilities (Hewer 2006).

2.6 Renewable Energy Alternatives and Climate Change

As a result of *climate change* by the 2070s, hydropower potential for the whole of Europe is expected to decline by 6%, translated into a 20–50% decrease around the Mediterranean, a 15–30% increase in northern and eastern Europe, and a stable hydropower pattern for western and central Europe (Lehner *et al.*, 2001).

Another possible conflict between *adaptation* and *mitigation* might arise over *water resources*. One obvious mitigation option is to shift to energy sources with low greenhouse gas emissions such as *small hydropower*. In regions, where hydropower potentials are still available, and also depending on the current and future water balance, this would increase the competition for water, especially if irrigation might be a feasible strategy to cope with climate change impacts on agriculture and the demand for cooling water by the power sector is also significant. This reconfirms the importance of integrated land and water management strategies to ensure the optimal allocation of scarce natural resources (land, water) and economic investments in climate change adaptation and mitigation and in fostering sustainable development. Hydropower leads to the key area of mitigation, energy sources and supply, and energy use in various economic sectors beyond land use, agriculture, and forestry.

The largest amount of construction work to counterbalance climate change impacts will be in water management and in coastal zones. The former involves hard measures in *flood protection* (dykes, dams, flood control reservoirs) and in coping with seasonal variations (storage reservoirs and inter-basin diversions), while the latter comprises coastal defense systems (embankment, dams, storm surge barriers).

Adaptation to changing hydrological regimes and water availability will also require continuous additional energy input. In water-scarce regions, the increasing reuse of waste water and the associated treatment, deep-well pumping, and especially large-scale desalination, would increase energy use in the water sector (Boutkan and Stikker 2004).

2.6.4 Biomass Energy

Overall 14% of the world's energy comes from *biomass*, primarily wood and charcoal, but also crop residue and even animal dung for cooking and some heating. This contributes to deforestation and the loss of topsoil in developing countries. Biofuel production is largely determined by the supply of moisture and the length of the growing season (Olesen and Bindi 2002). By the twenty-second century, land area devoted to biofuels may increase by a factor of two to three in all parts of Europe (Metzger *et al.*, 2004). Especially, in developing countries biomass is the major component of the national energy supply. Although biomass sources are widely available, they have low conversion efficiencies. This energy source is used especially for cooking and comfort and by burning it provides heat. The sun's radiation that conveys energy is exploited by the plants through *photosynthesis*, and consequently, even the remnants of plants are potential energy sources because they conserve historic solar energy until they perish either naturally after very long time spans or artificially by human beings or occasionally by forest fires. Only 0.1% of the solar incident energy is used by the photosynthesis process, but even this amount is ten times greater than the present day world energy consumption. Currently, living plants or remnants from the past are reservoirs of biomass that are a major source of energy for humanity in the future. However, biomass energy returns its energy to the atmosphere partly by respiration of the living plants and partly by oxidation of the carbon fixed by photosynthesis that is used to form fossil sediments which eventually transform to the fossil fuel types such as coal, oil, and natural gas. This argument shows that the living plants are the recipient media of incident solar radiation and they give rise to various types of fossil fuels.

Biofuel crops, increasingly an important source of energy, are being assessed for their critical role in adaptation to *climatic change* and mitigation of carbon emissions (Easterling *et al.*, 2007).

2.6.5 Wave Energy

Water covers almost two thirds of the earth, and thus, a large part of the sun's radiant energy that is not reflected back into space is absorbed by the water of the oceans. This absorbed energy warms the water, which, in turn, warms the air above and forms air currents caused by the differences in air temperature. These currents blow across the water, returning some energy to the water by generating wind *waves*, which travel across the oceans until they reach land where their remaining energy is expended on the shore.

The possibility of extracting energy from ocean waves has intrigued people for centuries. Although there are a few concepts over 100 years old, it is only in the past two decades that viable schemes have been proposed. Wave power generation is not a widely employed technology, and no commercial wave farm has yet been established. In the basic studies as well as in the design stages of a wave energy plant, the knowledge of the statistical characteristics of the local *wave climate* is essential, no matter whether physical or theoretical/numerical modeling methods are to be employed. This information may result from wave measurements, more or less sophisticated forecast models, or a combination of both, and usually takes the form of a set of representative sea states, each characterized by its frequency of occurrence and by a spectral distribution. Assessment of how turbo-generator design and the production of electrical energy are affected by the wave climate is very important. However, this may have a major economic impact, since if the equipment design is very much dependent on the wave climate, a new design has to be developed for each new site. This introduces extra costs and significantly limits the use of serial construction and fabrication methods.

Waves have an important effect in the planning and design of harbors, waterways, shore protection measures, coastal structures, and the other coastal works. Surface waves generally derive their energy from the wind. Waves in the ocean often have irregular shapes and variable propagation directions because they are under the influence of the wind. For operational studies, it is desired to forecast wave parameters in advance. Özger and Şen (2005) derived a modified average wave power formula by using perturbation methodology and a stochastic approach.

2.6.6 Hydrogen Energy

Hydrogen is the most abundant element on earth, however, less than 1% is present as molecular hydrogen gas H_2; the overwhelming part is chemically bound as H_2O in water and some is bound to liquid or gaseous hydrocarbons. It is thought that the heavy elements were, and still are, being built from hydrogen and helium. It has been estimated that hydrogen makes up more than 90% of all the atoms or 75% of the mass of the universe (Weast 1976). Combined with oxygen it generates water, and with carbon it makes different compounds such as methane, coal, and petroleum. Hydrogen exhibits the highest heating value of all chemical fuels. Furthermore, it is regenerative and environment friendly.

Solar radiation is abundant and its use is becoming more economic, but it is not harvested on large scale. This is due to the fact that it is difficult to store and move energy from ephemeral and intermittent sources such as the sun. In contrast, fossil fuels can be transported easily from remote areas to the exploitation sites. For the transportation of electric power, it is necessary to invest and currently spend money in large amounts. Under these circumstances of economic limitations, it is more rational to convert solar power to a gaseous form that is far cheaper to transport and easy to store. For this purpose, *hydrogen* is an almost completely clean-burning gas that can be used in place of petroleum, coal, or natural gas. Hydrogen does not release the carbon compounds that lead to *global warming*. In order to produce hydrogen, it is possible to run an electric current through water and this conversion process is known as *electrolysis*. After the production of hydrogen, it can be transported for any distance with virtually no energy loss. Transportation of gases such as hydrogen is less risky than any other form of energy, for instance, oil which is frequently spilled in tanker accidents, or during routine handling (Scott and Hafele 1990).

The ideal intermediary energy carrier should be storable, transportable, pollution-free, independent of primary resources, renewable, and applicable in many ways. These properties may be met by hydrogen when produced electrolytically using solar radiation, and hence, such a combination is referred to as the solar-hydrogen process. This is to say that transformation to hydrogen is one of the most promising methods of storing and transporting solar energy in large quantities and over long distances.

Among the many renewable energy alternatives, *solar-hydrogen energy* is regarded as the most ideal energy resource that can be exploited in the foreseeable future in large quantities. On the other hand, where conventional fuel sources are not available, especially in rural areas, solar energy can be used directly or indirectly by the transformation into hydrogen gas. The most important property of hydrogen is that it is the cleanest fuel, being non-toxic with virtually no environmental problems during its production, storage, and transportation. Combustion of hydrogen with oxygen produces virtually no pollution, except its combustion in air produces small amounts of nitrogen oxides. Solar-hydrogen energy through the use of hydrogen does not give rise to *acid rain*, greenhouse effects, *ozone* layer depletions, leaks, or spillages. It is possible to regard hydrogen after the treatment of water by solar energy as a synthetic fuel. In order to benefit from the unique properties of hydrogen, it must be produced by the use of a renewable source so that there will be no limitation or environmental pollution in the long run. Different methods have been evoked by using direct or indirect forms of solar energy for hydrogen production. These methods can be viewed under four different processes, namely:

1. Direct thermal decomposition or thermolysis
2. Thermo-chemical processes
3. Electrolysis
4. Photolysis

Large-scale hydrogen production has been obtained so far from the water electrolysis method, which can be used effectively in combination with photovoltaic cells. Hydrogen can be extracted directly from water by photolysis using solar radiation. Photolysis can be accomplished by photobiological systems, photochemical assemblies, or photoelectrochemical cells.

Hydrogen has been considered by many industrial countries as an environmentally clean energy source. In order to make further developments in the environmentally friendly solar-hydrogen energy source enhancement and research, the following main points must be considered:

1. It is necessary to invest in the research and development of hydrogen energy technologies
2. The technology should be made widely known
3. Appropriate industries should be established
4. A durable and environmentally compatible energy system based on the solar-hydrogen process should be initiated

Veziroğlu (1995) has suggested the following research points need to be addressed in the future to improve the prospects of solar-hydrogen energy:

1. Hydrogen production techniques coupled with solar and wind energy sources
2. Hydrogen transportation facilities through pipelines
3. Establishment and maintenance of hydrogen storage techniques
4. Development of hydrogen-fuelled vehicles such as busses, trucks, cars, *etc.*
5. Fuel cell applications for decentralized power generation and vehicles
6. Research and development on hydrogen hydrides for hydrogen storage and for air conditioning
7. Infrastructure development for solar-hydrogen energy
8. Economic considerations in any mass production
9. Environmental protection studies

On the other hand, possible demonstrations and/or pilot projects include the following alternatives:

1. Photovoltaic hydrogen production facility
2. Hydrogen production plants by wind farms
3. Hydro power plant with hydrogen off-peak generators
4. Hydrogen community
5. Hydrogen house
6. Hydrogen-powered vehicles

In order to achieve these goals, it is a prerequisite to have a data bank on the hydrogen energy industry, its products, specifications, and prices.

Another important and future promising technology for applying *solar photon energy* is the decomposition of water. This is referred to as the *Solar-Hydrogen Energy System* by Ohta (1979) and Justi (1987). Photolysis does not mean technically only water decomposition by photon energy, but also any photochemical reaction used to obtain the desired products.

2.7 Energy Units

In general, energy is defined as the ability to perform work. According to the first law of thermodynamics, the total sum of all the forms of energy in a closed system is constant. It is also referred to as the principle of energy conservation. In order to discuss quantitatively and comparatively various energy alternatives, it is necessary to bring them all to a common expression in terms of units of measurement.

The basic and physical unit of energy is the joule (J) which is based on the classic definition of work as the multiplication of force by distance. Hence, one joule is equivalent to the multiplication of one newton (N) of force by 1 m distance, and this definition gives J = Nm. The joule is named after the nineteenth century scientist, James Prescott Joule who demonstrated by experiments the equivalence of heat and work. Unfortunately, the joule is far too small a unit to be convenient for describing different resources of world energy supplies. It is, therefore, necessary to define greater versions such as the megajoule (MJ; 10^6 J), the gigajoule (GJ; 10^9 J), and the terajoule (TJ; 10^{12} J).

Another difficulty in practice with the joule is that oil producers measure the output of a well by barrels and coal producers by tons. Such different units require unification of the energy units by a common base. For instance, the *coal equivalent ton* (cet) is a basic unit which has been adopted by the United Nations. A commonly used value for the cet is 38.6×10^6 kJ. Likewise, it is also possible to define *oil equivalent ton* (oet) which is equal to 51×10^6 kJ. On the other hand, electrical energy is expressed, in general, in terms of kilowatt hours (kWh). It is, therefore, necessary to know the energy conversion factors between different energy units (Ohta 1979).

References

Anderson M (1992) Current status of wind farms in the UK. Renewable Energy System
Baldwin AH, Egnotovich MS, Clarke E (2001) Hydrologic change and vegetation of tidal fresh water marshes: field, greenhouse and seed-bank experiments. Wetlands 21:519–531
Boutkan E, Stikker A (2004) Enhanced water resource base for sustainable integrated water resource management. Nat Resour Forum 28:150–154
Bradley R, Fulmer R (2004) Energy: the master resource. Kendall/Hunt, Dubuque
Breslow PB, Sailor DJ (2002) Vulnerability of wind power resources to climate change in the continental United States. Renewable Energy 27:585–598
Burkett VR, Kusler J (2000) Climate change: potential impacts and interactions in wetlands of the United States. J Am Water Resour Assoc 36:313–320
Casola JH, Kay JE, Snover AK, Norheim RA, Whitely LC, Binder and Climate Impacts Group (2005) Climate impacts on Washington's hydropower, water supply, forests, fish, and agriculture. Prepared for King County (Washington) by the Climate Impacts Group (Center for Science in the Earth System, Joint Institute for the Study of the Atmosphere and Ocean, University of Washington, Seattle)
Clark A (1988) Wind farm location and environmental impact. Network for Alternative Technology and Technology Assessments C/O EEDU. The Open University, Milton Keynes, UK
Confalonieri U, Menne B, Akhtar R, Ebi KL, Hauengue M, Kovats RS, Revich B, Woodward A (2007) Human health. In: Parry ML, Canziani OF, Palutikof JP, van der Linden PJ, Hanson CE (eds) Climate change 2007: impacts, adaptation and vulnerability. Cambridge University Press, Cambridge, UK, pp 391–431
Easterling WE, Aggarwal PK, Batima P, Brander KM, Erda L, Howden SM, Kirilenko A, Morton J, Soussana J-F, Schmidhuber J, Tubiello FN (2007) Food, fiber and forest products. In: Parry ML, Canziani OF, Palutikof JP, van der Linden PJ, Hanson CE (eds) Climate change 2007: impacts, adaptation and vulnerability. Cambridge University Press, Cambridge, UK, pp 273–313
EWEA (1991) Time for action: wind energy in Europe. European Wind Energy Association
Harvey J (1982) θ-S relationships and water masses in the eastern North Atlantic. Deep Sea Res 29(8A):1021–1033
Hewer F (2006) Climate change and energy management: a scoping study on the impacts of climate change on the UK energy industry. UK Meteorological Office, Exeter
IPCC (2007) Parry ML, Canziani OF, Palutikof JP, van der Linden PJ, Hanson CE (eds) (2007) Climate change 2007: impacts, adaptation and vulnerability. Cambridge University Press, Cambridge, UK
Justi EW (1987) A solar-hydrogen energy system. Plenum, New York
Kundzewicz ZW, Mata LJ, Arnell NW, Döll P, Kabat P, Jiménez B, Miller KA, Oki T, Şen Z, Shiklomanov IA (2007) Freshwater resources and their management. In: Parry ML, Canziani OF, Palutikof JP, van der Linden PJ, Hanson CE (eds) Climate change 2007: impacts, adaptation and vulnerability. Cambridge University Press, Cambridge, UK, pp 173–210
Lehner B, Heinrichs T, Döll P, Alcamo J (2001) EuroWasser: model-based assessment of European water resources and hydrology in the face of global change. Kassel World Water Series 5, Center for Environmental Systems Research, University of Kassel, Germany
Metzger MJ, Leemans R, Schröter D, Cramer W, and the ATEAM consortium (2004) The ATEAM vulnerability mapping tool. Quantitative approaches in system analysis no. 27. Wageningen, C.T. de Witt Graduate School for Production Ecology and Resource Conservation, Wageningen, CD ROM
Monteith JL (1962) Attenuation of solar radiation: a climatological study. Q J R Meteorol Soc 88:508–521
Neuwirth F (1980) The estimation on global and sky radiation in Austria. Solar Energy 24:241
Ohta T (1979) Solar-hydrogen energy systems. Pergamon, New York

References

Olesen JE, Bindi M (2002) Consequences of climate change for European agricultural productivity, land use and policy. Eur J Agron 16:239–262

Özger M Şen Z (2005) Prediction of wave parameters by using fuzzy logic approach. Ocean Engineering 34:460–469

Pryor SC, Barthelmie RJ Kjellström E (2005) Potential climate change impact on wind energy resources in northern Europe: analyses using a regional climate model. Climate Dyn 25:815–835

Şahin AD (2004) Progress and recent trends in wind energy. Progr Energy Combustion Sci Int Rev J 30:501–543

Scott DS, Hafele W (1990) The coming hydrogen age: preventing world climatic disruption. Int J Hydrogen Energy 15:727–737

Şen Z (1995) Applied hydrogeology for scientists and engineers. CRC Lewis, Boca Raton

Şen Z (2003) A short physical note on a new wind power formulation. Renewable Energy 28:2379–2382

Şen Z (2005) Batmayan güneşlerimiz. (Our suns without sunset) (in Turkish) Bilim Serisi. Altın Burç Yayınları

Sun G, McNulty SG, Amatya DM, Skaggs RW, Swift LW, Shepard P, Riekerk H (2002) A comparison of watershed hydrology of coastal forested wetlands and the mountainous uplands in the Southern US. J Hydrol 263:92–104

Trenberth KE, Jones PD, Ambenje PG, Bojariu R, Easterling DR, Klein AMG, Parker DE, Renwick JA, Rahimzadeh F, Rusticucci MM, Soden BJ, Zhai PM (2007) Observations: surface and atmospheric change. In: Solomon S, Qin D, Manning M, Chen Z, Marquis M, Averyt KB, Tignor M, Miller HL (eds) Climate change 2007: the physical science basis. Cambridge University Press, Cambridge, UK, pp 235–336

Troen I, Peterson EL (1989) European wind atlas. Riso National Laboratory, Roskilde

Veziroğlu TN (1995) International Centre for Hydrogen Energy Technologies. Feasibility study. Clean Energy Research Institute, University of Miami, Coral Gables

Voisin N, Hamlet AF, Graham LP, Pierce DW, Barnett TP, Lettenmaier DP (2006) The role of climate forecasts in western U.S. power planning. J Appl Meteorol 45:653–673

Wark K, Warner CF (1981) Air pollution. Its origin and control. Harper Collins, New York

Weast RC (1976) Handbook of chemistry and physics, CRC Press, Boca Raton

Wilbanks TJ, Romero Lankao P, Bao M, Berkhout F, Cairncross S, Ceron J-P, Kapshe M, Muir-Wood R, Zapata-Marti R (2007) Industry, settlement and society. In: Parry ML, Canziani OF, Palutikof JP, van der Linden PJ, Hanson CE (eds) Climate change 2007: impacts, adaptation and vulnerability. Cambridge University Press, Cambridge, UK, pp 357–390

Chapter 3
Solar Radiation Deterministic Models

3.1 General

Solar radiation emission from the sun into every corner of space appears in the form of electromagnetic waves that carry energy at the speed of light. The solar radiation is absorbed, reflected, or diffused by solid particles in any location of space and especially by the earth, which depends on its arrival for many activities such as weather, climate, agriculture, and socio-economic movement. Depending on the geometry of the earth, its distance from the sun, geographical location of any point on the earth, astronomical coordinates, and the composition of the atmosphere, the incoming irradiation at any given point takes different shapes. A significant fraction of the solar radiation is absorbed and reflected back into space through atmospheric events and consequently the solar energy balance of the earth remains the same.

This chapter provides the basic astronomical variables and their definitions and uses in the calculation of solar radiation (energy) assessment. These basic concepts, definitions, and derived astronomical equations furnish the foundations of solar energy evaluation at any given location. They are deterministic in the sense that there is no uncertainty about the effect of weather events, which will be taken into consideration in the next three chapters.

3.2 The Sun

The sun has played a dominant role since time immemorial for different natural activities in the universe at large and in the earth in particular for the formation of fossil and renewable energy sources. It will continue to do so until the end of the earth's remaining life, which is predicted to be about 5×10^9 years. Deposited fossil fuels, in the form of coal, that are used through combustion are expected to last for approximately the next 300 years at the most, and from then onward human beings will be left with renewable energy resources only.

The diameter of the sun is $R = 1.39 \times 10^6$ km. The sun is an internal energy generator and distributor for other planets such as the earth. It is estimated that 90% of the energy is generated in the region between 0 and $0.23R$, which contains 40% of the sun's mass. The core temperature varies between 8×10^6 K and 40×10^6 K and the density is estimated at about 100 times that of water. At a distance $0.7R$ from the center the temperature drops to about 130,000 K where the density is about $70 \, \text{kg/m}^3$. The space from $0.7R$ to $1.0R$ is known as the *convective zone* with a temperature of about 5000 K and the density is about $10^{-5} \, \text{kg/m}^3$.

The observed surface of the sun is composed of irregular convection cells with dimensions of about 1000–3000 km and with a cell life time of a few minutes. Small dark areas on the solar surface are referred to as pores and have the same order of magnitude as the convective cells; larger dark areas are sunspots of various sizes. The outer layer of the convective zone is the photosphere with a density of about 10^{-4} that of air at sea level. It is essentially opaque as the gases are strongly ionized and able to absorb and emit a continuous spectrum of radiation. The *photosphere* is the source of most solar radiation. The recessing layer is above the photosphere and is made up of cooler gases several hundred kilometers deep. Surrounding this layer is the *chromosphere* with a depth of about 10,000 km. It is a gaseous layer with temperatures somewhat higher than that of the photosphere but with lower density. Still further out is the cornea, which is a region of very low density and very high temperature (about 10^6 K). Solar radiation is the composite result of the abovementioned several layers.

An account of the earth's energy sources and demand cannot be regarded as complete without a discussion of the sun, the solar system, and the place of the earth within this system. In general, the sun supplies the energy absorbed in the short term by the earth's atmosphere and oceans, but in the long term by the lithosphere where the *fossil fuels* are embedded. Conversion of some of the sun's energy into thermal energy derives the general atmospheric circulation (Becquerel 1839). A small portion of this energy in the atmosphere appears in the form of the kinetic energy of the winds, which in turn drive the ocean circulations. Some of the solar energy is intercepted by plants and is transformed by *photosynthesis* into biomass. In turn, a large portion of this is ultimately converted into heat energy by chemical oxidation within the bodies of animals and by the decomposition and burning of vegetable matter. On the other hand, a very small proportion of the photosynthetic process produces organic sediments, which may eventually be transformed into fossil fuels. It is estimated that the solar radiation intercepted by the earth in 10 days is equivalent to the heat that would be released by the combustion of all known reserves of fossil fuels on earth.

Until the rise of modern nuclear physics, the source of the sun's energy was not known, but it is now clear that the solar interior is a nuclear furnace that releases energy in much the same way as man-made thermonuclear explosions. It is now obvious through spectroscopic measurements of sunlight reaching the earth from the photosphere layer of the sun that the solar mass is composed predominantly of the two lightest elements, hydrogen, H, which makes up about 70% of the mass, and helium, He, about 27%; and the remaining 3% of solar matter is made up of all the

3.2 The Sun

other 90 or so elements (McAlester 1983). The origin of solar radiation received on the earth is the conversion of H into He through *solar fusion*. Theoretical considerations show that, at the temperatures and pressures of the solar interior, He is steadily being produced from lighter H as four nuclei unite to form one nucleus of helium as presented in Fig. 3.1. During such a conversion, single H nuclei (proton) made unstable by heat and pressure, first combine to form double H nuclei; these then unite with a third H nucleus to form ^{3}He, with a release of *electromagnetic energy*.

The sun is a big ball of *plasma* composed primarily of H and He and small amounts of other atoms or elements. Plasma is a state of matter where the electrons are separated from the nuclei because the temperature is so high and accordingly the kinetic energies of nuclei and electrons are also high. Protons are converted into He nuclei plus energy by the process of fusion. As schematically shown in Fig. 3.2 nuclei are composed of nucleons that come in two forms as protons and neutrons with positive and no charges, respectively.

This reaction is extremely exothermal and the free energy per He nuclei is 25.5 eV or 1.5×10^8 (kcal/g). The mass of four protons, 4×1.00723, is greater than the mass of the produced He nucleus 4.00151 by 0.02741 mass units. This small excess of matter is converted directly to *electromagnetic radiation* and is

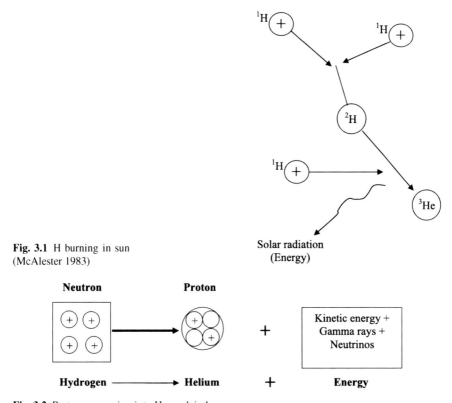

Fig. 3.1 H burning in sun (McAlester 1983)

Fig. 3.2 Proton conversion into He nuclei plus energy

the unlimited source of *solar energy*. The source of almost all renewable energy is the enormous fusion reactor in the sun which converts H into He at the rate of 4×10^6 tonnes per second. The theoretical predictions show that the conversion of four H atoms (*i. e.*, four protons) into He using carbon nuclei as a catalyst will last about 10^{11} years before the H is exhausted. The energy generated in the core of the sun must be transferred toward its surface for radiation into the space. Protons are converted into He nuclei and because the mass of the He nucleus is less than the mass of the four protons, the difference in mass (around 5×10^9 kg/second) is converted into energy, which is transferred to the surface where electromagnetic radiation and some particles are emitted into space; this is known as the solar wind.

It is well known by now that the planets, dust, and gases of the *solar system* that orbit around the enormous central sun contain 99.9% of the mass of the system and provide the gravitational attraction that holds it together. The average density of the sun is slightly greater than of water at $1.4 \, \text{g/cm}^3$. One of the reasons for sun's low density is that it is composed predominantly of H, which is the lightest element. Its massive interior is made up of matter held in a gaseous state by enormously high temperatures. Consequently, in smaller quantities, gases at such extreme temperatures would rapidly expand and dissipate. The emitted energy of the sun is 3.8×10^{26} W and it arises from the *thermonuclear fusion* of H into He at temperatures around 1.5×10^6 K in the core of the sun, which is given by the following chemical equation (Şen 2004) and it is comparable with Fig. 3.2:

$$4 {}^1_1 \text{H} \rightarrow {}^4_3 \text{He} + 2\beta + \text{energy}(26.7 \, \text{MeV})$$

In the core of the sun, the dominant element is He (65% by mass) and the H content is reduced to 35% by mass as a direct result of consumption in the fusion reactions. It is estimated that the remaining H in the sun's core is sufficient to maintain the sun at its present luminosity and size for another 4×10^9 years. There is a high-pressure gradient between the core of the sun and its perimeter and this is balanced by the gravitational attraction of the sun's mass. The energy released by the thermonuclear reaction is transported by energetic photons, but, because of the strong adsorption by the peripheral gases, most of these photons do not penetrate the surface. In all regions of the *electromagnetic spectrum*, the outer layers of the sun continuously lose energy by radiation emission into space in all directions. Consequently, a large temperature gradient exists between the core and the outer parts of the sun.

The sun radiates electromagnetic energy in terms of photons which are light particles. Almost one third of this incident energy on the earth is reflected back, but rest is absorbed and is, eventually, retransmitted to deep space in terms of long-wave infrared radiation. Today, the earth radiates just as much energy as it receives and sits in a stable energy balance at a temperature suitable for life on the earth. In fact, solar radiation is in the form of white light and it spreads over a wider spectrum of wavelengths from the short-wave infrared to ultraviolet. The wavelength distribution is directly dependent on the temperature of the sun's surface.

The total power that is incident on the earth's surface from the sun every year is 1.73×10^{14} kW and this is equivalent to 1.5×10^{18} kWh annually, which is equivalent to 1.9×10^{14} coal equivalent tons (cet). Compared to the annual world consumption of almost 10^{10} cet, this is a very huge and unappreciable amount. It is approximately 10,000 times greater than that which is consumed on the earth annually. In engineering terms, this energy is considered to be uniformly spread all over the world's surface and, hence, the amount that falls on one square meter at noon time is about 1 kW in the tropical regions. The amount of solar power available per unit area is known as irradiance or *radiant-flux density*. This solar power density varies with latitude, elevation, and season of the year in addition to time in a particular day (see Sect. 3.7). Most of the developing countries lie within the tropical belt of the world where there are high solar power densities and, consequently, they want to exploit this source in the most beneficial ways. On the other hand, about 80% of the world's population lives between latitudes 35° N and 35° S. These regions receive the sun's radiation for almost 3000–4000 h/year. In solar power density terms, this is equivalent to around 2000 kWh/year, which is 0.25 cet/year. Additionally, in these low latitude regions, seasonal sunlight hour changes are not significant. This means that these areas receive the sun's radiation almost uniformly throughout the whole year. Apart from the solar radiation, the sunlight also carries energy. It is possible to split the light into three overlapping groups:

1. Photovoltaic (PV) group: produces electricity directly from the sun's light
2. Photochemical (PC) group: produces electricity or light and gaseous fuels by means of non-living chemical processes
3. Photobiological (PB) group: produces food (animal and human fuel) and gaseous fuels by means of living organisms or plants

The last two groups also share the term *"photosynthesis"*, which means literally the building (synthesizing) by light.

3.3 Electromagnetic (EM) Spectrum

All solid, liquid, and gaseous matter is no more than a vibrating cosmic dance of energy. Matter is perceived by human beings in a three-dimensional form with structure, density, color, and sound. Density makes the matter solid, liquid or gaseous and, in addition, the movement of its atoms and molecules gives rise to the sensations of heat and cold. The interaction of matter with the area of the electromagnetic (EM) spectrum that is known as light gives it color, perceived through the eyes. However, if one takes a step inward, it can be observed that matter is composed of large and small molecules. Each atom, until the advent of modern physics, was considered to consist of a nucleus of positively charged protons and zero-charged neutrons, with a number of "shells" of orbiting, negatively charged electrons. With these particles, the H and He atoms are shown in Fig. 3.3. In modern physics, the subatomic particles are considered as wave packets, as electromagnetic force fields, and as energy relationships.

They have "spin" and they rotate about the axis of their movement. They have no "oscillation," like an ultra-high-speed pendulum. By spinning and oscillating, they

Fig. 3.3 a H atom. b He atom

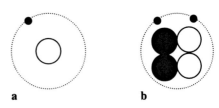

move around relative to each other in three dimensions. They also have an "electrical charge" and a "magnetic moment" and, therefore, an EM field. Radiation consists of atomic or subatomic particles, such as electrons, and/or EM energy waves, such as heat, light, radio and television signals, infrared, X-rays, gamma rays, *etc.*

Particle and wave are two forms of light and EM radiation. The former has localized mass in space and can have charge in addition to other properties. Two particles cannot occupy the same space at the same time. On the other hand, wave has no mass, is spread out over space at the speed of light, and obeys the superposition principle which means two or more waves can occupy the same space at the same time. For instance, EM waves consist of electric and magnetic fields, which are perpendicular to each other and perpendicular to the direction of travel as shown in Fig. 3.4.

The oscillating field planes of electric and magnetic waves are perpendicular to each other, *i. e.*, when the electric field E and magnetic field H_m are in the yz-plane, respectively, the propagation direction is along the x axis. Solar radiation EM waves travel with the speed of light and cover the distance between the sun and the earth in about 8 min. EM radiation from the sun is described by its wavelength, λ (distance from peak to peak of the wave), and frequency, f (number of cycles per second). As the wave moves past a location, the frequency is also expressed as the number

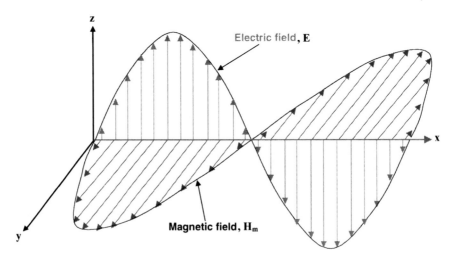

Fig. 3.4 EM waves

3.3 Electromagnetic (EM) Spectrum

of crests (peaks) per second. Wavelength and frequency (see Fig. 3.5) are related through the speed of light, c, as

$$c = \lambda f. \tag{3.1}$$

If either the wavelength or the frequency is known then the other can be calculated since the speed of light is a constant.

The *solar energy spectrum* contains wavelengths that are too long to be seen by the naked eye (the infrared) and also wavelengths that are too short to be visible (the ultraviolet). The spectral distribution of the solar radiation in W/m^2 per micrometer of wavelength gives the power per unit area between the wavelengths λ and $\lambda + 1$, where λ is measured in micrometers. The solar spectrum is roughly equivalent to a perfect black body at a temperature of 5800 K. After the combined effects of water vapor, dust, and adsorption by various molecules in the air, certain frequencies are strongly absorbed and as a result the spectrum received by the earth's surface is modified as shown in Fig. 3.6. The area under the curve gives the total power per square meter radiated by a surface at the specified temperature.

A blackbody is a perfect absorber or emitter of EM radiation. The intensity or energy of irradiation emitted per wavelength (or frequency) depends only on the temperature of the body and not on the type of material or atoms. So a blackbody spectrum curve can be generated for a specific temperature, with the peak of the curve shifting to shorter wavelengths (higher frequency) for higher temperatures. For instance, a blue flame is hotter than an orange one and objects at higher temperatures emit more radiation at all wavelengths, so the curves have similar shapes nested within one another as in Fig. 3.6. Notice that the peak of the curve for the sun is in the visible range and it is interesting that human eyes are most sensitive to yellow-green light.

As it can be seen from the same figure the maximum solar irradiance is at the wavelength $\lambda = 0.5\,\mu\text{m}$ which is in the region of the visible solar radiation from $\lambda = 0.4 - 0.7\,\mu\text{m}$. It also shows a standard spectral irradiance curve which has been compiled based on extensive measurements. The earth receives its radiation from the sun at short wavelength around a peak of $0.5\,\mu\text{m}$, whereas it radiates to space at a much lower wavelength around a peak value of $10\,\mu\text{m}$, which is well into the in-

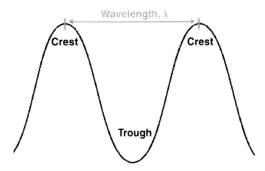

Fig. 3.5 Wave features

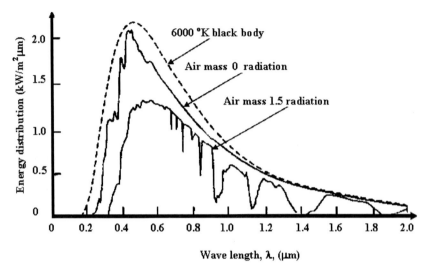

Fig. 3.6 Solar spectrum

frared. The relationship between the wavelength $\lambda_{\max}$, which is the power radiated at a maximum, and the body temperature, T, is given as Wien's law (Collares-Pereira and Rabl 1979; Frochlich and Werhli 1981) which reads as

$$\lambda_{\max} T = 3 \times 10^{-3} \text{ mK} . \quad (3.2)$$

EM waves show particle properties as photons and, in particular, they behave as if they were made up of packets of energy having an energy E, which is related to frequency f as

$$E = hf , \quad (3.3)$$

where h is the Plank constant, h $= 6.626 \times 10^{-34}$ J.s. The EM spectrum is the range of radiation from very short wavelengths (high frequency) to very long wavelengths (low frequency), as in Fig. 3.7. The subsections of the spectrum are labeled by how the radiation is produced and detected, but there is overlap between the neighboring ranges. At the atomic level, EM waves come in units as photons and a high frequency corresponds to high-energy photons.

The range of the visible spectrum is very small with red light having a longer wavelength (7×10^{-7} m) than blue light (4×10^{-7} m). In nature any rainbow is a familiar example of a few color mixtures from the spectrum, whereas white light is just a superposition (mixture) of all the colors. There are detectors for the whole range of EM radiation and, for instance, with an infrared detector it is possible to see objects in the dark.

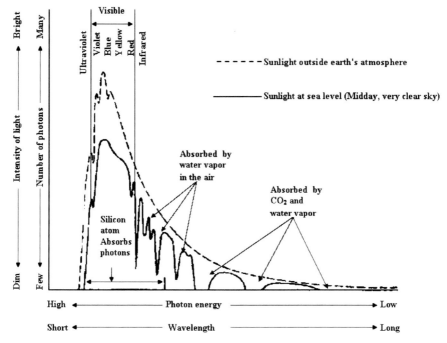

Fig. 3.7 EM spectrum

3.4 Energy Balance of the Earth

The earth radiates the same amount of energy into space as the amount of EM energy absorbed from the sun (Fig. 3.8). Hence, in the long term the energy balance of the earth is essentially zero, except for the small amount of geothermal energy generated by radioactive decay. If the in and out energies are not balanced then the earth is expected to increase or to decrease in temperature and radiate more or less energy into space to establish another balance level. At this point, it is useful to remember the present day *global warming* and *climate change* phenomena. Clear-sky shortwave solar radiation varies in response to altitude and elevation, surface gradient (slope), and orientation (aspect), as well as position relative to neighboring surfaces.

The sun's radiation (solar) energy first interacts with the atmosphere and then reaches to the earth's surface (Fig. 3.9). Incoming solar radiation of 100 units (100%) is shared by cloud and surface reflections and the rest is absorbed by the atmosphere and the earth's surface.

The atmospheric absorption accounts for about two thirds of the incoming irradiation and it is primarily due to water vapor and to a lesser degree by CO_2 that exist in the atmospheric composition (Chap. 2). Absorbed EM radiation is converted into thermal energy at absorption locations. The earth's surface has a comparatively lower temperature, typically between 270 K and 320 K, and hence radiates at longer

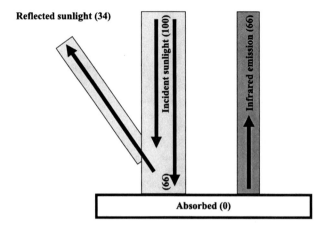

Fig. 3.8 Long-term earth energy balance

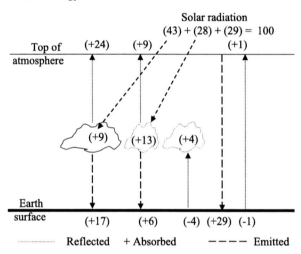

Fig. 3.9 Terrestrial solar radiation according to (Şen 2004)

wavelengths that do not appear appreciably in the spectrum of Fig. 3.7. This energy drives weather events in terms of evaporation and transportation of heat from the equator to the poles (Hadley cell) and additionally provides energy for wind and currents in the ocean with some absorption and storage in plants as *photosynthesis*. Some of this energy is radiated back to space (clear skies) as infrared radiation and the rest is absorbed in the atmosphere. Some of the infrared radiation goes back into space and the rest is re-radiated back to earth. Hence, clear nights are cooler than cloudy nights because the nighttime radiation into space has a temperature of 3 K. As the solar radiation reaches the upper boundary of earth's atmosphere, the light starts to scatter depending on the cloud cover and the atmospheric composition (Hay 1984). A proportion of the scattered light comes to earth as diffuse radiation. The term *"sunshine"* implies not the diffuse but the direct solar radiation

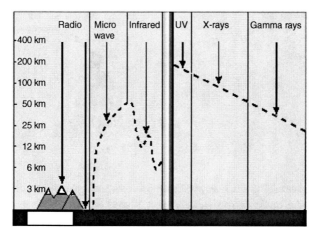

Fig. 3.10 Visible and radio waves reach the surface

(*solar beam*) that comes straight from the sun. On a clear day, direct radiation can approach a power density of $1000\,W/m^2$, which is known as *solar power density* for *solar collector* testing purposes. The atmosphere is largely transparent to visible and radio wavelengths, but absorbs radiation at other wavelengths (Fig. 3.10). This figure shows the EM wave recipient change by altitude. It is obvious that of all the EM waves radio waves are receivable at the lowest altitude.

3.5 Earth Motion

Earth's orbital movement around the sun affects the climate, solar radiation, and temporal variations. The total amount of solar radiation reaching the earth's surface can vary due to changes in the sun's output, such as those associated with earth's axis tilt, wobble, and orbital trace. Orbital oscillations can also result in different parts of the earth getting more or less sunlight even when the total amount reaching the planet remains constant, which is similar to the way the tilt in the earth's axis produces the hemispheric seasons. Due to the variations in the earth's orbital movements around the sun there are very slow climate cycles. The earth's rotation about its own axis and revolution around the sun is very involved and has the combination of three movements at any time. These are as follows:

1. *Tilt* (Fig. 3.11): Due to the axis tilt the sun's motion across the sky changes during each year. The tilt of the earth changes cyclically between 21°45' and 24°15' with a cyclic period of 42,000 years. A large tilt warms the poles, which means more solar radiation input and causes smaller temperature differences in the summer hemisphere.
   The tilt of the earth's rotational axis with respect to the sun is called *obliquity*, which is defined as the angle between the earth's orbit and the plane of the

Fig. 3.11 Changes in tilt

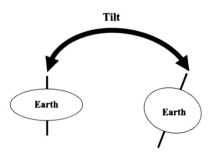

earth's equator. The tilt is toward the sun in the summer hemisphere and away from the sun in the winter hemisphere. Winter occurs in November–February (May–August) in the northern (southern) hemisphere because of the tilt of the earth's axis (Fig. 3.12). Additionally, diurnal variations are also effective due to the day and night sequence.

As a result of the earth's rotation around a tilted axis, surprisingly the polar region receives more radiation in the summer than at the equator. An important feature is the absence of seasons at the tropics and the extremes of a six-month summer and a six-month winter at the poles (Dunn 1986).

2. *Wobble* (see Fig. 3.13): As it rotates, the earth wobbles on its axis like a spinning top, with a period of 23,000 years. These wobbles affect the amount of solar energy the earth receives and where that energy is deposited. This in turn affects the climate, introducing regular cycles with periods of up to 100,000 years.

As the earth wobbles, its axis sweeps out an imaginary cone in space. This is known as *precession* of the earth's axis.

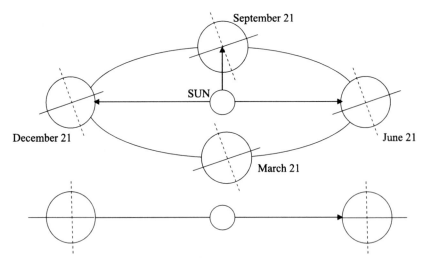

Fig. 3.12 Seasons are due to the tilt of earth's axis

3.5 Earth Motion

Fig. 3.13 Changes in wobble

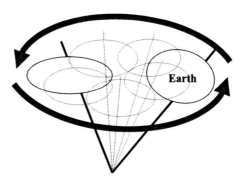

3. *Orbit* (see Fig. 3.14): The earth moves rotationally once a year around the sun on an *elliptical* orbit that is almost in the form of a circle. In the mean time it spins on its axis once a day. Its closest position to the sun appears on 1 January and furthest position on 1 July. During the orbital movement the radius from the sun determines how much solar radiation (energy) is available and the tilt of the spin axis as 23°30' relative to the orbital plane causes the seasons with different solar radiation rates. The shape of the earth's path around the sun ranges from a nearly perfect circle to a more elliptic shape over the 100,000-year cycle (*eccentricity*). This means that the sun not the centroid of the earth's orbit causes the distance from the earth to the sun to vary. The amount of solar energy received by the earth is greatest when the earth is nearest to the sun. This factor combined with the tilt of the earth's axis causes seasonal climate changes, which are out of phase in each hemisphere. For instance, northern hemisphere winters are currently milder and summers cooler than normal. The opposite situation, colder winters and hotter summers, is now occurring in the southern hemisphere.

These small variations in the earth–sun geometry change how much sunlight each hemisphere receives during the earth's year-long trek around the sun, where in the orbit (the time of year) the seasons occur, and how extreme the seasonal changes are. The amount of solar energy per area depends on the angle of the surface in relation to the sun (Fig. 3.15). For example, on 21 December there is no sunlight at any location above the Arctic Circle, which corresponds to 66°30' N latitude.

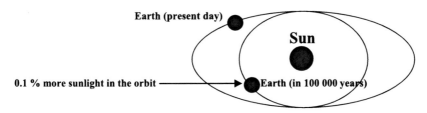

Fig. 3.14 Changes in orbit

Fig. 3.15 Earth's axis tilt affect on incident radiation per area

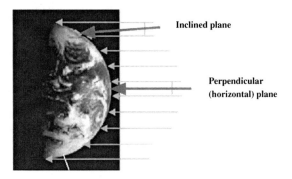

In this figure, it is summer at the south pole and there is sunlight 24 h a day, while at the north pole there is no sunlight (see Sect. 3.12.3). In the northern hemisphere it is winter, because the same amount of incident radiation is spread over a larger area. Notice the parallel solar radiation lines on the plane at the top and at the equator, which has the same length as the parallel lines between rays. Of course, the same solar energy will impinge on both plates, but the one at the top will have more surface area and hence its solar energy absorption will be comparatively smaller than the other plate at the equator. This implies that the more the angle (*zenith angle*) between the normal of the plate and the solar beam, the less is the solar energy generation (see Sect. 3.12).

This tremendous amount of solar energy radiates into space from the surface of the sun with a power of 3.8×10^{23} kW. Solar energy is referred to as renewable and/or *sustainable energy* because it will be available as long as the sun continues to shine. The energy from the sunshine, EM radiation, is referred to as *insolation*. The earth intercepts only a very small portion of this power, since the projected area of the earth as seen from the sun is very small. At the top of the atmosphere, the power intercepted by the earth is 173×10^9 W, which is equivalent to $1360 \, \text{W/m}^2$. On a clear day, at the surface of the earth, the solar radiation is about $1000 - 1200 \, \text{W/m}^2$, on a plane perpendicular to the sun's beam depending on the elevation and the amount of haze in the atmosphere.

In order to appreciate the arrival of solar radiation on the earth's surface, it is very helpful to simplify the situation as shown in Fig. 3.15 where the earth is represented as a sphere. This implies that at the equator a horizontal surface at that point immediately under the sun would receive $1360 \, \text{W/m}^2$. Along the same longitude but at different latitudes, the horizontal surface receives less solar radiation from the equator toward the polar region. If the earth rotates around the vertical axis to the earth–sun plane, then any point on the earth's surface receives the same amount of radiation throughout the year. However, the earth rotates around an axis which is inclined to the earth–sun plane, and therefore, the same point receives different amounts of solar radiation on different days and times in a day throughout the year. Hence, the seasons start to play a role in the incident solar radiation variation.

3.6 Solar Radiation

Solar radiation from the sun after traveling in space enters the atmosphere at the space–atmosphere interface, where the ionization layer of the atmosphere ends. Afterwards, a certain amount of solar radiation or photons are absorbed by the atmosphere, clouds, and particles in the atmosphere, a certain amount is reflected back into the space, and a certain amount is absorbed by the earth's surface. The earth's surface also reflects a certain amount of energy by radiation at different wavelengths due to the earth's surface temperature. About 50% of the total solar radiation remains in the atmosphere and earth's surface. The detailed percentages can be seen in Fig. 3.9. The earth's rotation around its axis produces hourly variations in power intensities at a given location on the ground during the daytime and results in complete shading during the nighttime.

The presence of the atmosphere and associated climate effects both attenuate and change the nature of the solar energy resource. The combination of reflection, absorption (filtering), refraction, and scattering result in highly dynamic radiation levels at any given location on the earth. As a result of the cloud cover and scattering sunlight, the radiation received at any point is both direct (or beam) and diffuse (or scattered).

After the solar radiation enters the earth's atmosphere, it is partially scattered and partially absorbed. The scattered radiation is called *diffuse radiation*. Again, a portion of this diffuse radiation goes back to space and a portion reaches the ground. Solar radiation reaches the earth's surface in three different ways as direct, diffuse, and reflected irradiations as in Fig. 3.16.

The quantity of solar radiation reaching any particular part of the earth's surface is determined by the position of the point, time of year, atmospheric diffusion, cloud cover, shape of the surface, and reflectivity of the surface.

However, in hilly and mountainous terrains, the distribution of slopes has major effects on surface climate and radiation amounts. Surface radiation may change widely according to the frequency and optical thickness of clouds, and modeling these cloud properties successfully is important for treatment of the surface energy balance (Chap. 4).

Direct solar radiation is that which travels in a straight line from the sun to the earth's surface. Clear-sky day values are measured at many localities in the world. To model this would require knowledge of intensities and direction at different times of the day. Direct radiation as the name implies is the amount of solar radiation received at any place on the earth directly from the sun without any disturbances. In practical terms, this is the radiation which creates sharp shadows of the subjects. There is no interference by dust, gas, and cloud or any other intermediate material on the direct solar radiation. Direct radiation is practically adsorbed by some inter-mediator and then this inter-mediator itself radiates EM waves similar to the main source which is the sun. Direct solar radiation can be further reflected and dispersed across the surface of the earth or back into the atmosphere. On the other hand, the radiation arriving on the ground directly in line from the sun is called direct or *beam radiation* (Fig. 3.16a). Beam radiation is the solar radiation received from the sun

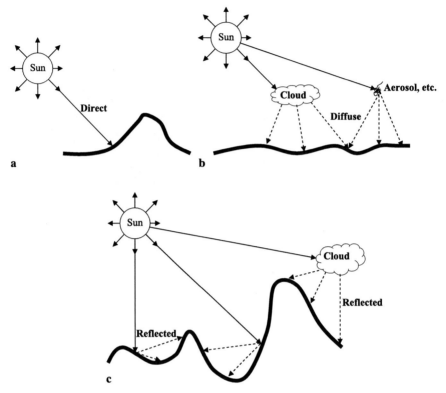

Fig. 3.16 a–c. Solar radiation paths. **a** Direct. **b** Diffuse. **c** Reflected

without scatter by the atmosphere. It is referred to as direct solar radiation. This is actually the photon stream in space and has a speed of 3,000,000 km/s.

Passing through the atmosphere, the solar beam undergoes wavelength- and direction-dependent adsorption and scattering by atmospheric gases, aerosols, and cloud droplets. The scattered radiation reaching the earth's surface is referred to as diffuse radiation (Fig. 3.16b). Diffuse radiation is first intercepted by the constituents of the air such as water vapor, CO_2, dust, aerosols, clouds, *etc.*, and then released as scattered radiation in many directions. This is the main reason why diffuse radiation scattering in all directions and being close to the earth's surface as a source does not give rise to sharp shadows. When the solar radiation in the form of an *electromagnetic* wave hits a particle, a part of the incident energy is scattered in all directions and it is called diffuse radiation. All small or large particles in nature scatter radiation. *Diffuse radiation* is scattered out of the solar beam by gases (Rayleigh scattering) and by aerosols (which include dust particles, as well as sulfate particles, soot, sea salt particles, pollen, *etc.*). Reflected radiation is mainly reflected from the terrain and is therefore more important in mountainous areas. Direct shortwave radiation is the most important component of global radiation because it contributes the most to the energy balance and also the other components depend

3.6 Solar Radiation

on it, either directly or indirectly (Kondratyev 1965). If the particles are spherical and much smaller than the wavelength of the incident radiation, it is referred to as *Rayleigh scattering*. In Rayleigh scattering, the scattering process is identical in forward and backward directions with a minimum scattering in between. When the particle size is of the order of incident radiation wavelength, the solution of the wave equation becomes formidable. In this case, the scattering is called *Mie's scattering* and more energy is scattered in a forward than in a backward direction. On any clear day, the diffuse component from the Rayleigh and aerosol scattering is about 10–30% of the total incident radiation, whereas when the solar beam passes through a cloud essentially all the surface radiation is diffuse. This radiation consists of solar photons arriving from all directions of the sky, with intensities depending on the incoming direction. Diffuse radiation occurs when small particles and gas molecules diffuse part of the incoming solar radiation in random directions without any alteration in the wavelength of the electromagnetic energy.

Solar energy modeling requires knowledge of surface reflectance and shape, and a means of modeling any dispersal. Albedo is a measure of how much radiation is reflected by a surface (Table 2.1). When the *albedo* is 1.0 all radiation is reflected; none is absorbed. When the albedo is 0.0 no radiation is reflected; it is all absorbed (Graves 1998). A significant proportion of direct solar radiation striking a surface is *reflected*, particularly from snow and clouds. What proportion of the reflected radiation strikes another surface is not known.

Diffuse radiation occurs when small particles and gas molecules diffuse part of the incoming solar radiation in random directions without any alteration in the wavelength of the electromagnetic energy. Diffuse cloud radiation would require modeling of clouds, which was considered impossible to do and would have been variable from day to day. It appears to only contribute a minor part to radiation energies from above the mid-visible through to the infrared spectrum, but can contribute up to 40% of the radiation energy from the mid-visible through to the mid-ultraviolet spectrum (Barbour *et al.*, 1978).

Total (*global*) solar radiation is the sum of the beam and the diffuse solar radiation on a surface. The most common measurements of solar radiation are total radiation on a horizontal surface, hereafter referred to as global radiation on the surface. The total solar radiation is sometimes used to indicate quantities integrated over all wavelengths of the solar spectrum. The sun's total energy is composed of 7% ultraviolet (UV) radiation, 47% visible radiation, and 46% infrared (heat) radiation. UV radiation causes many materials to degrade and it is significantly filtered out by the ozone layer in the upper atmosphere.

The total global radiation at the earth's surface consists of both short- and long-wave radiation. Short-wave radiation may be absorbed by terrestrial bodies and cloud cover and re-emitted as long-wave radiation. The short-wave radiation reaching the surface of the earth may be direct, diffuse, or reflected (Fig. 3.16).

Global radiation at a location is roughly proportional to direct solar radiation, and varies with the geometry of the receiving surface. The other components, such as diffuse radiation, vary only slightly from slope to slope within a small area and the variations can be linked to slope gradient (Kondratyev 1965; Williams *et al.*, 1972).

In fact, diffuse radiation comprises less than 16% of the total irradiance at visible wavelengths in the green and red region (Dubayah 1992), rising to 30% for blue. The flux of *clear-sky* diffuse radiation varies with slope orientation in much the same way as the flux of direct solar radiation, hence preserving the spatial variability in total radiation (Dubayah et al., 1989).

3.6.1 Irradiation Path

Sun-born solar radiation rays travel toward the earth and on their way they encounter many molecules, the number of which is dependent on the distance traveled, d, which is called the path distance (or slant path). If the distance-dependent density is $\rho(d)$ then the actual optical mass, m_a, can be expressed as

$$m_a = \int_0^\infty \rho(d)\,\mathrm{d}d\,. \tag{3.4}$$

This is valid for *monochromatic radiation* because the refraction along the traveled path depends on the wavelength. If the sun is at its zenith at a location then Eq. 3.4 can be written as

$$m_z = \int_0^\infty \rho(z)\,\mathrm{d}z\,, \tag{3.5}$$

where z is the distance on the zenith direction (Fig. 3.17).

If the path is different than the zenith direction then its optical path can be expressed as the ratio, m, of Eq. 3.4 to Eq. 3.5,

$$m = \frac{\int_0^\infty \rho(d)\,\mathrm{d}d}{\int_0^\infty \rho(z)\,\mathrm{d}z}\,. \tag{3.6}$$

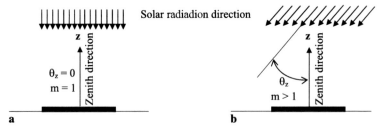

Fig. 3.17 a,b. Solar radiation paths

3.6 Solar Radiation

This is referred to as the relative *optical air mass*. On the basis of the assumptions that the earth does not have any eccentricity and the troposphere is completely homogeneous and free of any *aerosol* or water vapor, then the relative optical mass, m, in any direction with an angle of θ_z from the zenith can be written simply as

$$m = m_a \sec\theta_z = \frac{m_a}{\cos\theta_z}. \tag{3.7}$$

If the assumptions are applied to an actual case, this expression yields errors of up to 25% at $\theta_z = 60°$, which decrease to 10% at $\theta_z = 85°$ (Iqbal 1986). At sea level, $m = 1$ when the sun is at the zenith and $m = 2$ for $\theta_z = 60°$.

Air mass as defined in Eq. 3.7 is a useful quantity in dealing with atmospheric effects. It indicates the relative distance that light must travel through the atmosphere to a given location. There is no attenuation effect in the space outside the atmosphere, the air mass is regarded as equal to zero and as equal to one when the sun is directly overhead. However, an air mass value of 1.5 is considered more representative of average terrestrial conditions and it is commonly used as a reference condition in rating photovoltaic modules and arrays (Fig. 3.18).

The two main factors affecting the air mass ratio are the direction of the path and the local altitude. The path's direction is described in terms of its zenith angle, θ_z, which is the angle between the path and the zenith position directly overhead. The adjustment in air mass for local altitude is made in terms of the local atmospheric pressure, p, and is defined as

$$m = \frac{p}{p_0} m_0, \tag{3.8}$$

where p is the local pressure and m_0 and p_0 are the corresponding air mass and pressure at sea level. Equation 3.7 is valid only for zenith angles less than 70° (Kreith

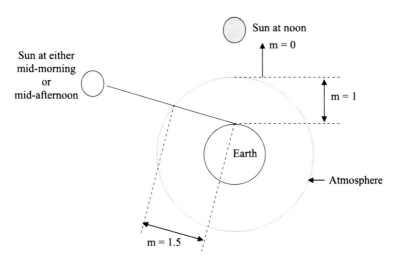

Fig. 3.18 Sun's angle and distance through atmosphere

and Kreider 1978). Otherwise, the secant approximation under-estimates solar energy because atmospheric refraction and the curvature of the earth have not been accounted for. Frouin *et al.* (1989) have suggested the use of the following:

$$m = \left[\cos\theta_z + 0.15\,(93.885 - \theta_z)^{-1.253}\right]^{-1}. \tag{3.9}$$

Kreith and Kreider (1978) and Cartwright (1993) have suggested the use of the following relationship, where the model requires the calculation of air mass ratio as

$$m = \left[1229 + (614\sin\alpha)^2\right]^{1/2} - 614\sin\alpha. \tag{3.10}$$

It is obvious then that the relative proportion of direct to diffuse radiation depends on the location, season of the year, elevation from the mean sea level, and time of day. On a clear day, the diffuse component will be about 10–20% of the total radiation but during an overcast day it may reach up to 100%. This point implies, practically, that in the solar radiation and energy calculations, weather and meteorological conditions in addition to the astronomical implications must be taken into consideration. On the other hand, throughout the year the diffuse solar radiation amount is smaller in the equatorial and tropical regions than the sub-polar and polar regions of the world. The instantaneous total radiation can vary considerably through the day depending on the cloud cover, dust concentration, humidity, *etc.*

3.7 Solar Constant

The sun's radiation is subject to many absorbing, diffusing, and reflecting effects within the earth's atmosphere which is about 10 km average thick and, therefore, it is necessary to know the power density, *i. e.*, watts per meter per minute on the earth's outer atmosphere and at right angles to the incident radiation. The density defined in this manner is referred to as the *solar constant*. The solar constant and the associated spectrum immediately outside the earth's atmosphere are determined solely by the nature of the radiating sun and the distance between the earth and the sun.

Earth receives virtually all of its energy from space in the form of solar EM radiation. Its total heat content does not change significantly with time, indicating a close overall balance between absorbed solar radiation and the diffuse stream of low-temperature, thermal radiation emitted by the planet. The radiance at the mean solar distance – the solar constant – is about 1360 W/m^2 (Monteith 1962). At the mean earth–sun distance the sun subtends an angle of 32'. The radiation emitted by the sun and its spatial relationship to the earth result in a nearly fixed intensity of solar radiation outside the earth's atmosphere. The solar constant, I_0 (W/m^2), is the energy from the sun per unit time per unit area of surface perpendicular to the direction of the propagation of the radiation. The measurements made with a variety of instruments in separate experimental programs resulted as $I_0 = 1353$ W/m^2 with an estimated error of $\pm 1.5\%$. The World Radiation Center

3.8 Solar Radiation Calculation

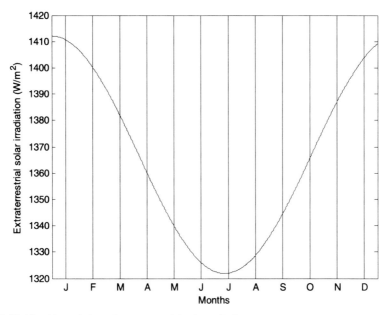

Fig. 3.19 Monthly variation of extraterrestrial solar radiation

has adopted a value of 1367 W/m² with an uncertainty of 1%. The most updated solar constant is $I_0 = 1367$ W/m², which is equivalent to $I_0 = 1.960$ cal/cm² min or 432 Btu/ft²h or 4.921 MJ/m²h.

Iqbal (1986) gives more detailed information on the solar constant. As the distance between the sun and the earth changes during the whole year the value of the solar constant changes also during the year as in Fig. 3.19.

The best value of the solar constant available at present is $I_0 = 1360$ W/m² (Frochlich and Werhli 1981). Whereas the solar constant is a measure of solar power density outside the earth's atmosphere, terrestrial applications of photovoltaic (Chap. 7) devices are complicated by the following two variables that must be taken into consideration:

1. Atmospheric effects (Chap. 2)
2. Geometric effects, including the earth's rotation about its tilted axis and its orbital revolution around the sun (Sect. 3.11)

3.8 Solar Radiation Calculation

Solar irradiance, I (W/m²), is the rate at which radiant energy is incident on a unit surface. The incident energy per unit surface is found by integration of irradiance over a specified time, usually an hour or a day. *Insolation* is a term specifically for solar energy irradiation on surfaces of any orientation.

There are two dimensions to the *energy flux* due to the energy of photons and the energy itself. Specialists in "solar energy" think in terms of an integrated expression over a certain time interval and have the dimension of energy, J, which is "insolation" as the integrated "irradiance".

In general, modeling the solar radiation arriving at the top of the atmosphere can simply be considered as the product of the solar constant I_0 and the astronomical factor $f(R)$ of annual average 1.0, proportional to R^{-2} (inverse distance square), where R is the distance of the earth from the sun. However, for the modeling of solar radiation at the earth's land surfaces, it is usually adequate to assume that the diffuse radiation is *isotropic* (the same intensity in all directions from the sky). Diffuse sky irradiance under cloud-free conditions may be estimated by assuming an isotropic sky and calculating the proportion of the sky seen from a point [that is using the equivalent of the view-shed operation in GIS (Dubayah and Rich 1995)]. Under cloudy or partly cloudy conditions, diffuse radiation is anisotropic which may be explicitly modeled, but in practice this is computationally expensive to achieve as the diffuse radiation from different portions of the sky must be calculated. In order to calculate actual solar flux, field data from *pyranometers* (which measure actual incoming solar flux at a station), atmospheric optical data, or atmospheric profiling (sounding) must be used.

If I is the intensity of radiation arriving at the ground surface from a given direction, then the amount incident per unit surface area along the zenith direction is

$$I_z = I \cos \theta_z \,, \tag{3.11}$$

where θ_z is the *azimuth angle* between the normal to the surface and the direction of the beam (see Fig. 3.20). In the simplest modeling efforts land is assumed to be horizontal.

The solar radiation varies according to the orbital variations. If I_T is the total solar radiation output from the sun at all frequencies then at a distance R from the sun's center, the flux of the radiation will be the same assuming that the radiation

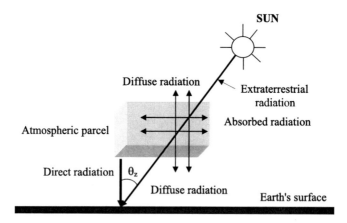

Fig. 3.20 Surface solar radiation

3.8 Solar Radiation Calculation

is equal in all directions. If the radiation flux per unit area at a distance R is represented by $Q(R)$, then the total radiation is equal to $4\pi R^2 Q(R)$. Hence, it is possible to write that

$$I_T = 4\pi R^2 Q(R) \qquad (3.12)$$

or

$$Q(R) = \frac{I_T}{4\pi R^2}. \qquad (3.13)$$

The earth is approximately 150×10^6 km away from the sun, hence Eq. 3.13 yields approximately that the total solar output is 3.8×10^{26} W. Of course, the radiation incident on a spherical planet is not equal to the solar constant of that planet. The earth intercepts a disk of radiation from the sun with area πR^2, where R is the radius of the earth. Since, the surface area of the earth is $4\pi R^2$, the amount of solar radiation per unit area on a spherical planet becomes

$$\frac{\pi R^2 Q(R)}{4\pi R^2} = \frac{Q(R)}{4}. \qquad (3.14)$$

Consequently, the average radiation on the earth's surface can be calculated as $1360/4 = 340\,\text{W/m}^2$. All these calculations assume that the earth is perfectly spherical without any atmosphere and revolves on a circular orbit without *eccentricity*. Of course, these simplifications must be used in practical applications.

The driving force for the atmosphere is the absorption of solar energy at the earth's surface. Over time scales which are long compared to those controlling the redistribution of energy, the earth–atmosphere system is in thermal equilibrium. The absorption of solar radiation, at visible wavelengths as *short-wave* (SW) radiation, must be balanced by the emission to space of infrared or *long-wave* (LW) radiation by the planet's surface and atmosphere. A simple balance of SW and LW radiations leads to an equivalent blackbody temperature for the earth as $T = 255$ K. This is some 30 K colder than the global mean surface temperature, $T_s \approx 288$ K. The difference between these two temperatures follows from the *greenhouse effect* which results from the different ways the atmosphere processes SW and LW radiations. Although transparent to SW radiation (wavelength $\approx 1\,\mu\text{m}$), the same atmosphere is almost opaque to LW radiation (wavelength $\approx 10\,\mu\text{m}$) re-emitted by the planet's surface. By trapping radiant energy that must eventually be rejected into space, the atmosphere's capacity elevates the surface temperature beyond what it would be in the absence of an atmosphere.

The change in *extraterrestrial* solar radiation can be calculated by taking into account the astronomical facts according to the following formula:

$$I = I_0 \left[1 + 0.033 \cos\left(\frac{360 N_d}{365}\right)\right], \qquad (3.15)$$

where N_d is the number of the day corresponding to a given date. It is defined as the number of days elapsed in a given year up to a particular date starting from 1 on

1 January to 365 on 31 December. On the other hand, solar radiation is attenuated as it passes through the atmosphere and, in a simplified case, may be estimated according to an exponential decrease by using Bouger's Law (Kreith and Kreider 1978) as

$$I = I_0 e^{-km} \quad (3.16)$$

where it is assumed that the sky is clear, I and I_0 are the terrestrial and extraterrestrial intensities of beam radiation, k is an absorption constant, and m is the air mass ratio.

3.8.1 Estimation of Clear-Sky Radiation

As the solar radiation passes through the earth's atmosphere it is modified due to the following reasons:

1. Absorption by different gases in the atmosphere
2. Molecular (or Rayleigh) scattering by the permanent gases
3. Aerosol (Mie) scattering due to particulates

Absorption by atmospheric molecules is a selective process that converts incoming energy to heat, and is mainly due to water, oxygen, ozone, and carbon dioxide. Equations describing the absorption effects are given by Spencer (1972). A number of other gases absorb radiation but their effects are relatively minor and for most practical purposes can be ignored (Forster 1984).

Atmospheric scattering can be either due to molecules of atmospheric gases or due to smoke, haze, and fumes (Richards 1993). Molecular scattering is considered to have a dependence inversely proportional to the fourth power of the wavelength of radiation, *i. e.*, λ^{-4}. Thus the molecular scattering at 0.5 mm (visible blue) will be 16 times greater than at 1.0 mm (near-infrared). As the primary constituents of the atmosphere and the thickness of the atmosphere remain essentially constant under clear-sky conditions, molecular scattering can be considered constant for a particular wavelength. Aerosol scattering, on the other hand, is not constant and depends on the size and vertical distribution of the particulates. It has been suggested (Monteith and Unsworth 1990) that a $\lambda^{-1.3}$ dependence can be used for continental regions. In an ideal clear atmosphere *Rayleigh* scattering is the only mechanism present (Richards 1993) and it accounts for the blueness of the sky. The effects of the atmosphere in absorbing and scattering solar radiation are variable with time as atmospheric conditions and the air mass ratio change. Atmospheric transmittance, τ, values vary with location and elevation between 0 and 1. According to Gates (1980) at very high elevations with extremely clear air τ may be as high as 0.8, while for a clear sky with high turbidity it may be as low as 0.4.

As shown in Figs. 3.16 and 3.20, the solar radiation during its travel through the atmosphere toward the earth surface meets various phenomena, including scatter, absorption, reflection, diffusion, meteorological conditions, and air mass, which change with time. It is useful to define a standard atmosphere "clear" sky and cal-

3.8 Solar Radiation Calculation

culate the hourly and daily radiation that would be received on a horizontal surface under these standard conditions. Hottel (1976) has presented a method of estimating the beam radiation transmitted through a clear atmosphere and he introduced four climate types as in Table 3.1. The atmospheric transmittance for beam radiation, τ, is given in an exponentially decreasing form depending on the altitude, A, and zenith angle as

$$\tau = a + b\exp\left(-\frac{c}{\cos\theta_z}\right), \quad (3.17)$$

where the estimations of constants a, b, and c for the standard atmosphere with 23 km visibility are given for altitudes less then 2.5 km by (Kreith and Kreider 1978)

$$a = 0.4237 - 0.00821\,(6 - A)^2, \quad (3.18)$$
$$b = 0.5055 - 0.005958\,(6.5 - A)^2, \quad (3.19)$$

and

$$c = 0.2711 - 0.01858\,(2.5 - A)^2, \quad (3.20)$$

where A is the altitude of the observer in kilometers. The correction factors (r_a, r_b and r_c) are given for four climate types (Table 3.1).

Kreith and Kreider (1978) have described the atmospheric transmittance for beam radiation by the empirical relationship

$$\tau = 0.56\left(e^{-0.65\,m} + e^{-0.095\,m}\right). \quad (3.21)$$

The constants account for attenuation of radiation by the different factors discussed above. Since scattering is wavelength dependent, the coefficients represent an average scattering over all wavelengths. This relationship gives the atmospheric transmittance for clear skies to within 3% accuracy (Kreith and Kreider 1978) and the relationship has also been used by Cartwright (1993). The atmospheric transmittance in Eq. 3.21 can be replaced by site-specific values, if they are available, and hence the solar radiation on a horizontal plane can be estimated as

$$I = I_0 \tau. \quad (3.22)$$

Table 3.1 Correction factors

Climate type	r_a	r_b	r_c
Tropical	0.95	0.98	1.02
Mid-latitude summer	0.97	0.99	1.02
Sub-artic summer	0.99	0.99	1.01
Mid-latitude winter	1.03	1.01	1.00

3.9 Solar Parameters

Solar radiation and energy calculations require some geometric and time quantities concerning the sun position relative to the earth and any point on the earth. It is also necessary to know the relation between the local standard time and the solar time.

3.9.1 Earth's Eccentricity

It is desirable to have the distance and the earth's eccentricity in mathematical forms for simple calculations. Although a number of such forms are available of varying complexities, it is better to have simple and manageable expressions such as the one suggested by Spencer (1972), who gave the eccentricity, ε, correction factor of the earth's orbit as

$$\varepsilon = 1.00011 + 0.034221\cos\Gamma + 0.001280\sin\Gamma \\ + 0.000719\cos 2\Gamma + 0.000077\sin 2\Gamma, \quad (3.23)$$

where day angle, Γ, in radians is given as

$$\Gamma = \frac{2\pi(N_d - 1)}{365}. \quad (3.24)$$

On the other hand, in terms of degrees one can write the day angle as

$$\Gamma = \frac{360(N_d - 1)}{365.242}. \quad (3.25)$$

Duffie and Backman (1991) suggested a simple approximation for ε as follows:

$$\varepsilon = 1 + 0.033\cos\left(\frac{2\pi N_d}{365}\right). \quad (3.26)$$

The use of this last expression instead of Eq. 3.23 does not make an appreciable difference. The average distance between the sun and the earth is $R = 150 \times 10^6$ km. Due to the eccentricity of the earth's orbit, the distance varies by 1.7%.

3.9.2 Solar Time

Solar time is based not only on the rotation of the earth about its axis but also on the earth's revolution around the sun during which the earth does not sweep equal areas on the ecliptic plane (see Fig. 3.21).

3.9 Solar Parameters

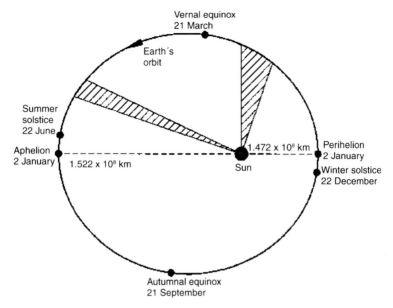

Fig. 3.21 Earth's orbit around the sun

3.9.2.1 Equation of Time

These combined movements, as already mentioned in Sect. 3.5, cause small discrepancies in the sun's appearance exactly over the local meridian daily. This discrepancy is the align of time, E_t, which is expressed by Spencer (1972) as

$$E_t = 229.18\,(0.000075 + 0.001868\cos\Gamma - 0.032077\sin\Gamma \\ -0.014615\cos 2\Gamma - 0.04089\sin 2\Gamma)\,. \tag{3.27}$$

The constant multiplication factor in front of the parenthesis on the right-hand side is for conversion from radians to minutes. It varies in length throughout the year due to the following factors:

1. The tilt of the earth's axis with respect to the plane of the ecliptic containing the respective centers of the sun and the earth
2. The angle swept out by the earth–sun vector during any given period of time, which depends upon the earth's position in its orbit (see Fig. 3.21).

The standard time (as recorded by clocks running at a constant speed) differs from the solar time. The difference between the standard time and solar time is defined as the align of time, E_t, which may be obtained as expressed by Woolf (1968):

$$E_t = 0.1236\sin\Gamma - 0.0043\cos\Gamma + 0.1538\sin 2\Gamma + 0.0609\cos 2\Gamma\,, \tag{3.28}$$

Table 3.2 Harmonic coefficients

k	A_k (10^3) (h)	B_k (10^3) (h)
0	0.2087	0.00000
1	9.2869	−122.29000
2	−52.2580	−156.98000
3	−1.3077	−5.16020
4	−2.1867	−2.98230
5	−1.5100	−0.23463

where Γ must be substituted from Eqs. 3.24 or 3.25. E_t may also be obtained more precisely as presented by Lamm (1981) in the form of harmonic components as

$$E_t = \sum_{k=0}^{5} \left[A_k \cos \frac{2\pi k N}{365.25} + B_k \sin \frac{2\pi k N}{365.25} \right], \quad (3.29)$$

where N is the day in the 4-year cycle starting after the leap year. The harmonic coefficients are given in Table 3.2.

3.9.2.2 Apparent Solar Time (AST)

Most meteorological measurements are recorded in terms of local standard time. In many solar energy calculations, it is necessary to obtain irradiation, wind, and temperature data for the same instant. It is, therefore, necessary to compute local apparent time, which is also called the true solar time. Solar time is the time to be used in all solar geometry calculations. It is necessary to apply the corrections due to the difference between the local longitude, L_{loc}, and the longitude of the standard time meridian, L_{stm}. The apparent time, L_{at}, can be calculated by considering the standard time, L_{st}, according to Iqbal (1986) as

$$L_{at} = L_{st} \pm 4(L_{stm} − L_{loc}) + E. \quad (3.30)$$

In this expression + (−) sign is taken in degrees toward the west (east) of the 0° meridian (longitude), which passes through Greenwich in the UK. All terms in the above equation are to be expressed in hours.

3.9.3 Useful Angles

The basic angles that are necessary in the definition of the geographic locations are latitude, θ, and longitude, ϕ. The latitude is the angular distance measured along a meridian from the equator (north or south) to a point on the earth's surface. Any location towards the north (south) has positive (negative) latitude with maximum degrees as +90 (−90) at the north (south) pole. On the other hand, longitude is the

3.9 Solar Parameters

angular distance measured from the prime (solar noon) meridian through Greenwich, UK, west or east to a point on the earth's surface. Any location west (east) of the prime meridian is positive (negative) location (see Fig. 3.22).

The two significant positions of the sun are height above the horizon, which is referred to as the solar altitude at noon and it changes by 47° from 21 June to 21 December. At the equinoxes on 21 March and 21 September, at noon the sun is directly overhead with 90° at the equator and sunrise (sunset) is due east (west) for all locations on the earth. The winter (summer) solstice corresponds to dates that the sun reaches it highest (lowest) positions at solar noon in each hemisphere. On 21 December the sun is directly overhead at 23°30' south latitude and on 21 June it is directly overhead at the same degree north latitude. These two latitudes are called the "tropic of Capricorn" and "tropic of Cancer," respectively.

The position of the sun can be calculated for any location and any time as shown in Fig. 3.23. The position of the sun is given by two angles, which are altitude, α_A, and azimuth angle, γ. The altitude (or elevation) is the angle of the sun above the horizon and azimuth (or bearing) is the angle from north to the projection on the earth of the line to the sun. The solar position is symmetrical about solar noon (which is different than 12 noon local time). Irradiation fluctuates according to the weather and the sun's location in the sky. This location constantly changes throughout the day due to changes in both the sun's altitude (or elevation) angle and its azimuth angle. Figure 3.23 shows the two angles used to specify the sun's location in the sky.

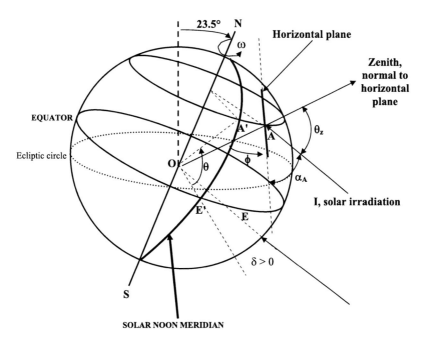

Fig. 3.22 Useful angles

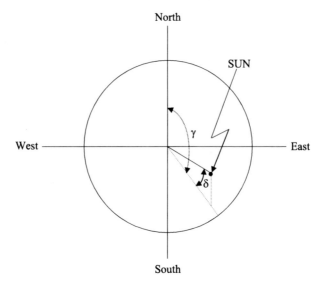

Fig. 3.23 Position of the sun by altitude and azimuth

The zenith angle, θ_z, is the angle between the vertical and the line connecting to the sun (the angle of incidence of beam radiation on a horizontal surface). Likewise, the angle between the horizontal and the line to the sun is the solar altitude angle, (the complement angle of the zenith angle), hence $\alpha_A + \theta_z = 90°$.

The angle between the earth–sun line and the equatorial plane is called the declination angle, δ, which changes with the date and it is independent of the location (see Fig. 3.22). The declination is maximum 23°45' on the summer/winter solstice and 0° on the equinoxes (Fig. 3.24).

The following accurate expression is considered for declination angle, δ, in radians and the eccentricity correction factor of the earth's orbit as defined above

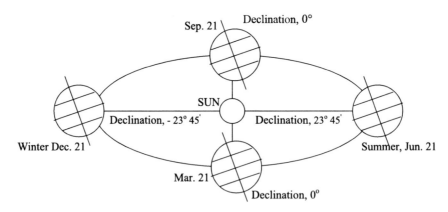

Fig. 3.24 The declination angles

(Spencer 1972):

$$\delta = 0.006918 - 0.399912\cos\Gamma + 0.07257\sin\Gamma - 0.006758\cos2\Gamma$$
$$+ 0.000907\sin\Gamma - 0.002697\cos3\Gamma + 0.00148\sin3\Gamma. \quad (3.31)$$

It is also possible to consider the following expressions for the approximate calculations of δ and E_0 (Iqbal 1986) as

$$\delta = 23.45\sin\left[\frac{360(284 + N_d)}{365}\right]. \quad (3.32)$$

As stated by Jain (1988) this expression estimates δ with a maximum error of 3' and Eq. 3.23 estimates ε with a maximum error of 0.0001.

Declination angle is considered to be positive when the earth–sun vector lies northward of the equatorial plane. Declination angle may also be defined as the angular position of the sun at noon with respect to the equatorial plane. It may be obtained as

$$\sin\delta = 0.39795\cos[0.98563(N_d - 1)], \quad (3.33)$$

where the cosine term is to be expressed in degrees and hence the arc sine term will be returned in radians (Kreider and Kreith 1981).

The hour angle, ω, is the angular distance that the earth rotates in a day, which is equal to $15°$ multiplied by the number of hours ($15 \times 24 = 360°$) from local solar noon (see Fig. 3.22). This is based on the nominal time, 24 h, required for the earth to rotate once i. e., $360°$. Values east (west) of due south (north), morning (evening) are positive (negative). Hence, the ω can be defined by

$$\omega = 15(12 - h), \quad (3.34)$$

where h is the current hour of the day.

The solar altitude is the vertical angle between the horizontal and the line connecting to the sun. At sunset (sunrise) altitude is $0°$ and $90°$ when the sun is at the zenith. The altitude relates to the latitude of the site, the declination angle, and the hour angle.

3.10 Solar Geometry

The geometric relationship between the sun and the earth can be described by the latitude of the site, the time of the year, the time of the day, the angle between the sun and the earth, and the altitude and azimuth angles of the sun. Geometric fundamentals, which are needed in solar radiation calculations, are presented in Fig. 3.25, including the beam of direct solar irradiance I reaching a point A on horizontal terrain.

The solar declination, δ, latitude, θ, hour angle (longitude), ϕ, and the earth's angular rotational velocity, ω, are the essential geometry and variables involved in the determination of the duration of daily irradiation, and energy input by solar radiation. O is the earth's center, and I is the vector of direct solar radiation reaching the

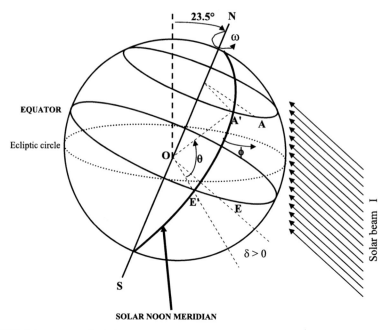

Fig. 3.25 Solar energy calculation geometry

earth's surface in a parallel beam. The point A is located uniquely by the latitude θ and hour angle ϕ (longitude). A positive (negative) longitude (hour angle) is measured counterclockwise (clockwise) from the solar noon meridian to the meridian containing the point A. The hour angle is depicted by the circular sector, A to A', on a plane parallel to the equatorial plane.

The simple relationship between the hour angle, ϕ, and the earth's rotational angular velocity, ω, is $\omega = d\phi/dt$, where approximately 2π radians is equal to 24 h in the counterclockwise direction in the northern hemisphere. Time $t = 0$ is chosen to correspond to solar noon (location at Greenwich, UK). The beam of direct solar radiation strikes perpendicular to a horizontal surface at point E' on the solar noon meridian and its latitude is called the solar declination (δ). Its range is approximately $-23.45° \leq \delta \leq 23.45°$ (Stacey 1992), being positive (negative) when it is the northern (southern) hemisphere. The solar declination is about $+23.45°(-23.45°)$ on the summer (winter) solstice in the northern (southern) hemisphere, and equals zero on the autumnal and vernal equinoxes.

3.10.1 Cartesian and Spherical Coordinate System

Daily irradiation duration calculation on a sloping terrain can be appreciated using a spherical coordinate system, which is very suitable for observing the passage from

3.10 Solar Geometry

a horizontal to a sloping surface. The mutually orthogonal unit directions $\mathbf{u}_r$, $\mathbf{u}_\theta$, and $\mathbf{u}_\phi$ in a spherical coordinate system are shown in Fig. 3.26. Hence, an oblique view can be obtained for the solar noon meridian and of the meridian containing the sloping surface point A of latitude θ and hour angle ϕ. Here, the latitude is positive (negative) for the northern (southern) hemisphere.

Geographically, the longitude and latitude (hour angle) can be expressed very conveniently as the components of a spherical coordinate system with its origin at the earth's center. Hence, any point on the earth will have geographically its latitude, θ, on the line that connects this point to the earth's center and perpendicular to this line at the same geographic point the longitude, ϕ, from the solar noon half meridian. The first axis in the spherical coordinate system is the radial line from the earth's center to the location of the point on the earth's surface; it falls on the radius of the earth and is denoted by r with its unit vector as $\mathbf{u}_r$. This is the direction of the horizontal plane normal vector. On the same point perpendicular to this axis there is the second axis of the spherical coordinate system that lies within the meridian and it is tangential to the earth's surface. The unit vector of this axis is from the point A

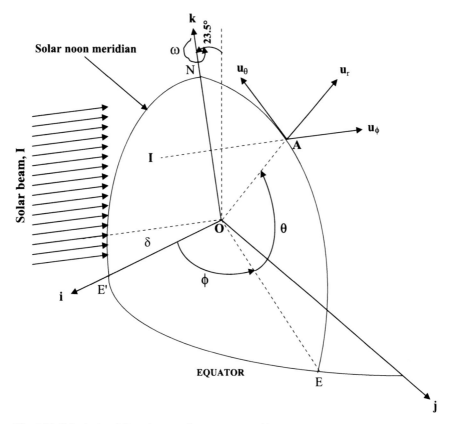

Fig. 3.26 Spherical and Cartesian coordinate system positions

toward the north (south) with positive (negative) angle values. Its unit vector is $\mathbf{u}_\theta$ as shown in Fig. 3.26. The completion of the spherical coordinate system requires the third axis perpendicular to the previous two axes and it is tangential at the same earth point to the latitude circle and its unit vector is $\mathbf{u}_\phi$.

On the other hand, there is another coordinate system in the form of Cartesian axes that go through the earth's center with the z axis directed toward the north along the earth's rotational axis with unit vector $\mathbf{k}$. The x axis with its unit vector $\mathbf{i}$, goes through the earth's center and it constitutes the intersection line between the equator plane and the solar noon half meridian plane. The third Cartesian coordinate axis is perpendicular to these two axes and has unit vector $\mathbf{j}$. The change of the geographic point on the earth changes the spherical system accordingly, but the Cartesian system remains as it is. In solar energy calculations, it is necessary to refer to the constant coordinate system, which is the defined as the Cartesian system and hence all the directions must be expressed in terms of $(\mathbf{i}, \mathbf{j}, \mathbf{k})$ unit vectors.

The unit vector $\mathbf{u}_r$ is radial outward at point A and is perpendicular to the horizontal plane tangential at point A where $\mathbf{u}_\theta$ is in the direction of increasing absolute value of the latitude and it is tangential to the meridian containing point A and $\mathbf{u}_\phi$ has the direction of increasing hour angle and in the mean time it is perpendicular to both $\mathbf{u}_\theta$ and $\mathbf{u}_r$. Hence, these unit vectors can be related to a Cartesian coordinate system unit direction vectors $\mathbf{i}$, $\mathbf{j}$, and $\mathbf{k}$ by considering the earth's center point O.

In Fig. 3.25, N indicates the north pole, and I is the direct solar radiation vector. The k axis coincides with the direction of the line segment O-N, which is part of the earth's rotation axis, and the other two unit vectors, i and j, fall on the equatorial plane. The $\mathbf{u}_r$, $\mathbf{u}_\theta$, and $\mathbf{u}_\Phi$ unit vectors can be expressed in terms of the latitude, hour angle, and the Cartesian unit vectors $\mathbf{i}$, $\mathbf{j}$, and $\mathbf{k}$, by considering from Fig. 3.26 the projections of spherical coordinates on the Cartesian coordinate system as follows:

$$\mathbf{u}_r = (\cos\theta \cos\phi)\mathbf{i} + (\cos\theta \sin\phi)\mathbf{j} + (\sin\theta)\mathbf{k}, \tag{3.35}$$

$$\mathbf{u}_\theta = (-\sin\theta \cos\phi)\mathbf{i} + (-\sin\theta \sin\phi)\mathbf{j} + (\cos\theta)\mathbf{k}, \tag{3.36}$$

and

$$\mathbf{u}_\phi = (-\sin\phi)\mathbf{i} + (\cos\phi)\mathbf{j}. \tag{3.37}$$

These equations are valid for horizontal planes with its normal vector that falls on to the $\mathbf{u}_r$ direction. Hence, there is no need for further calculation in order to define the position of the horizontal plane. It is convenient to remember at this point that, in the solar radiation and energy calculations, so far as the planes are concerned their positions are depicted with the normal vectors. For the sake of argument, let us define a plane with its three dimensions as the thickness, T_r, length, L_ϕ, and width, W_θ, where each subscript indicates the direction of each quantity. In other words, any horizontal plane is defined by its latitudinal width, longitudinal length, and radial thickness, as shown in Fig. 3.27.

This plane is horizontal in the sense that its normal direction coincides with the radial axis of the spherical coordinate system (Fig. 3.26). Hence, the rotation of this plane around the radial axis causes the plane to remain horizontal whatever the

3.10 Solar Geometry

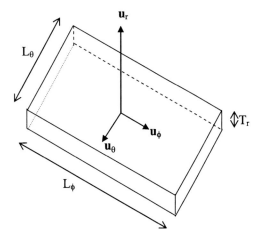

Fig. 3.27 Solar horizontal plane

rotation angle. On the other hand, the horizontal plane becomes inclined in one of the three rotations:

1. *Rotation around the ϕ axis:* It gives the plane a tilt angle, hence the new axis as r', θ', and ϕ' have the configuration in Fig. 3.28.

   On the other hand, in the case of a sloping surface a view perpendicular to the great circle containing the solar noon meridian appears as in Fig. 3.29.

   Such a two-dimensional view shows several of the geometric factors governing the solar radiation of a sloping surface where solar irradiance, I, falls at the base of a slope, which can be downward or upward from the horizontal plane tangential at A'. Herein, a positive (negative) sign to a downward (upward) slope is considered. The downward slope in this figure is rotated $\alpha°$, which is the critical angle, and if exceeded, would result in a shaded slope during solar noon. If the view in Fig. 3.25 represents the northern hemisphere summer solstice then the solar declination would be $\delta = 23.45°$, and day-long irradiation would be experienced at latitudes $66.55° \leq \theta° \leq 90°$. On the other hand, Fig. 3.30 is a graphical

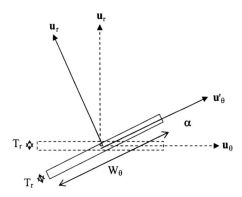

Fig. 3.28 ϕ axis rotation

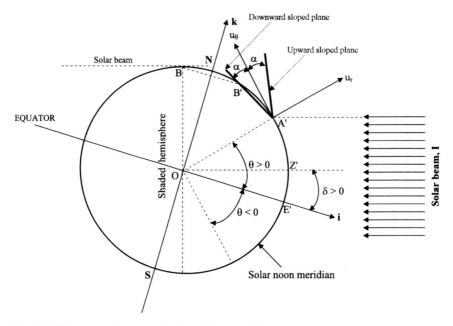

Fig. 3.29 View perpendicular to the plane of a great circle

summary of a rotation of the coordinate vectors $\mathbf{u}_r$, $\mathbf{u}_\theta$, and $\mathbf{u}_\phi$ exerted to achieve a downward (positive) slope. The third unit vector $\mathbf{u}_\phi = \mathbf{u}'_\phi$ is perpendicular onto the plane and serves as the axis of rotation in this instance. An upward (negative) slope would be achieved by making the rotation in a clockwise direction.

Here, α is the tilt angle or the slope angle, which is counted as positive toward the north as upward slope. The new position of the plane has $\mathbf{u}'_r$ as the new normal and $\mathbf{u}'_\theta$ axis perpendicular to it. This rotation will leave the unit vector of the ϕ axis the same, hence, $\mathbf{u}_\phi = \mathbf{u}'_\phi$. The relationship between ($\mathbf{u}'_r$ and $\mathbf{u}'_\theta$) and the original axes unit vectors can be written as

$$\mathbf{u}'_r = (\cos\alpha)\mathbf{u}_r - (\sin\alpha)\mathbf{u}_\theta, \qquad (3.38)$$
$$\mathbf{u}'_\theta = (\sin\alpha)\mathbf{u}_r + (\cos\alpha)\mathbf{u}_\theta, \qquad (3.39)$$

and

$$\mathbf{u}'_\phi = \mathbf{u}_\phi. \qquad (3.40)$$

Hence, the substitution of Eqs. 3.35 and 3.36 conveniently into the first two equations gives the inclined plane expressions with respect to longitude and latitude as follows:

$$\mathbf{u}'_r = (\cos\alpha\cos\theta + \sin\alpha\sin\theta)\cos\phi\mathbf{i} + (\cos\alpha\cos\theta + \sin\alpha\sin\theta)\sin\phi\mathbf{j}$$
$$- (\cos\alpha\sin - \sin\alpha\cos\theta)\mathbf{k}, \qquad (3.41)$$

3.10 Solar Geometry

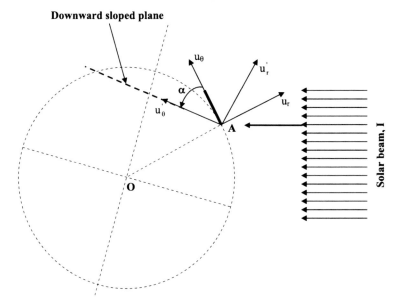

Fig. 3.30 Rotation of the spherical coordinate system to achieve a desired slope, α

$$\mathbf{u}'_\theta = (\sin\alpha\cos\theta - \cos\alpha\sin\theta)\cos\phi\mathbf{i} + (\sin\alpha\cos\theta - \cos\alpha\sin\theta)\sin\phi\mathbf{j}$$
$$+ (\sin\alpha\sin\theta + \cos\alpha\sin\theta)\mathbf{k}, \quad (3.42)$$

and

$$\mathbf{u}'_\phi = \mathbf{u}_\phi, \quad (3.43)$$

or more succinctly

$$\mathbf{u}'_r = \cos(\theta - \alpha)\cos\phi\mathbf{i} + \cos(\theta - \alpha)\sin\phi\mathbf{j} + \sin(\theta - \alpha)\mathbf{k}, \quad (3.44)$$
$$\mathbf{u}'_\theta = \sin(\theta - \alpha)\cos\phi\mathbf{i} + \sin(\theta - \alpha)\sin\phi\mathbf{j} + \cos(\theta + \alpha)\mathbf{k}, \quad (3.45)$$

and

$$\mathbf{u}'_\phi = \mathbf{u}_\phi. \quad (3.46)$$

2. *Rotation around the θ axis:* It gives to the plane an aspect angle of Ω (see Fig. 3.31). In this case, the θ axis remains the same and the plane can be defined with its new normal direction along the $\mathbf{u}''_r$ and perpendicular axis to it as $\mathbf{u}''_\phi$.
Similar to the previous case the Eqs. 3.41–3.43 remain the same except α will be replaced by Ω and finally the relevant expressions are expressed succinctly as

$$\mathbf{u}''_r = \cos(\theta - \Omega)\cos\phi\mathbf{i} + \cos(\theta - \Omega)\sin\phi\mathbf{j} + \sin(\theta - \Omega)\mathbf{k}, \quad (3.47)$$

Fig. 3.31 θ axis rotation

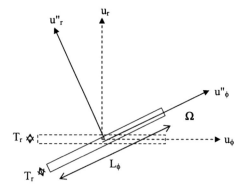

$$\mathbf{u}''_\phi = \sin(\theta - \Omega)\cos\phi \mathbf{i} + \sin(\theta - \Omega)\sin\phi \mathbf{j} + \cos(\theta + \Omega)\mathbf{k}, \quad (3.48)$$

and

$$\mathbf{u}''_\theta = \mathbf{u}_\theta. \quad (3.49)$$

3. *Two successive rotations, first around the ϕ axis and then subsequently around the θ axis:* In this manner both α and Ω angles will be effective as in Fig. 3.32. Figure 3.33 shows how a slope can be rotated $\Omega°$ west (east) of north to achieve aspects ranging from $0°$ west (east) of north to $180°$ west (east) of north. The accepted convention is to make Ω positive (negative) if the rotation is west (east) of north.

The rotations in Fig. 3.33 with respect to the unit vectors $\mathbf{e}'_\phi$ and $\mathbf{e}'_\theta$ (see also Fig. 3.32) lead to doubly rotated unit vectors $\mathbf{u}'''_r (= \mathbf{u}'_r)$, $\mathbf{u}'''_\theta$, and $\mathbf{u}'''_\phi$. These doubly rotated unit vectors provide the coordinate system with which to describe the geometry of a sloping surface in full generality, and they are given by the following equations:

$$\mathbf{u}'''_r = [\cos\alpha\cos\theta\cos\phi - \sin\alpha\cos\Omega\sin\theta\cos\phi + \sin\alpha\sin\Omega\sin\phi]\mathbf{i} +$$
$$[\cos\alpha\cos\theta\sin\phi - \sin\alpha\cos\Omega\sin\theta\sin\phi - \sin\alpha\sin\Omega\cos\phi]\mathbf{j} +, \quad (3.50)$$
$$[\cos\alpha\sin\theta + \sin\alpha\cos\Omega\cos\theta]\mathbf{k}$$

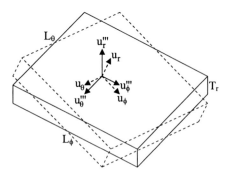

Fig. 3.32 ϕ and θ axes rotation

3.11 Zenith Angle Calculation

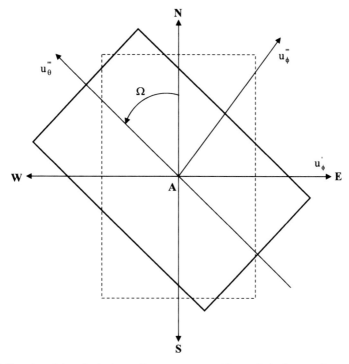

Fig. 3.33 Rotation of the spherical coordinate system to achieve a desired aspect Ω

$$\begin{aligned}\mathbf{u}_{\theta}''' =& [-\sin\alpha\cos\theta\cos\phi - \cos\alpha\cos\Omega\sin\theta\cos\phi + \sin\alpha\sin\Omega\sin\phi]\mathbf{i}+ \\ & -[\sin\alpha\cos\theta\sin\phi + \cos\alpha\cos\Omega\sin\theta\sin\phi + \cos\alpha\sin\Omega\cos\phi]\mathbf{j}+ \\ & [-\sin\alpha\sin\theta + \cos\alpha\cos\Omega\cos\theta]\mathbf{k}\end{aligned} \quad (3.51)$$

and

$$\begin{aligned}\mathbf{u}_{\phi}''' =& -[-\cos\Omega\sin\phi + \sin\Omega\sin\theta\cos\phi]\mathbf{i}+ \\ & [\cos\Omega\cos\phi - \sin\Omega\sin\theta\sin\phi]\mathbf{j}+ \\ & [\sin\Omega\cos\theta]\mathbf{k}\,.\end{aligned} \quad (3.52)$$

In the case of $\alpha = \Omega = 0$ the unit vectors in Eqs. 3.50–3.52 revert to those in Eqs. 3.35–3.37, respectively.

3.11 Zenith Angle Calculation

It is possible to calculate the angle by considering the scalar vector multiplication between the solar beam and the normal to the plane directions (Figs. 3.26, 3.33;

Eq. 3.50), which can be expressed as

$$\mathbf{I} \bullet \mathbf{u}_r''' = |\mathbf{I}| \cdot |\mathbf{u}_r'''| \cos\theta_z \,, \qquad (3.53)$$

where • indicates scalar multiplication. Equation 3.53 can be re-written as

$$\cos\theta_z = \frac{\mathbf{I} \bullet \mathbf{u}_r'''}{|\mathbf{I}| \cdot |\mathbf{u}_r'''|} \,. \qquad (3.54)$$

The absolute values on the right hand side are intensity of the vectors. From Fig. 3.26 one can express the solar radiation direction vector as follows:

$$\mathbf{I} = (-\cos\delta)\mathbf{i} + (-\sin\delta)\mathbf{k} \,. \qquad (3.55)$$

The scalar multiplication in the numerator of Eq. 3.54 is equal to the multiplication of the corresponding components of the two vectors. By considering that $\mathbf{i} \bullet \mathbf{i} = \mathbf{j} \bullet \mathbf{j} = \mathbf{k} \bullet \mathbf{k} = 1$ and $\mathbf{i} \bullet \mathbf{j} = \mathbf{i} \bullet \mathbf{k} = \mathbf{j} \bullet \mathbf{k} = 0$. The substitution of Eqs. 3.54 and 3.55 into Eq. 3.47 yields

$$\cos\theta_z = (\cos\delta \sin\alpha \cos\Omega \sin\theta - \cos\delta \cos\alpha \cos\theta) \cos\phi$$
$$- \cos\delta \sin\alpha \sin\Omega \sin\theta - \sin\delta \cos\alpha \sin\theta - \sin\delta \sin\alpha \cos\Omega \cos\theta \,. \quad (3.56)$$

This is the general expression and it is possible to deduce special case solutions. For instance, in the case of the horizontal plane $\alpha = \Omega = 0$ and the resulting equation is

$$\cos\theta_z = -\cos\delta \cos\theta \cos\phi - \sin\delta \sin\theta \,. \qquad (3.57)$$

The same expression can be reached by the scalar multiplication of the sun beam and $\mathbf{u}_r$ vectors from Eqs. 3.33 and 3.55. This implies that the solar radiation direction is in the opposite of r (zenith) direction, and therefore, one can write actually that

$$\cos\theta_z = \cos\delta \cos\theta \cos\phi + \sin\delta \sin\theta \,. \qquad (3.58)$$

On the other hand, if the plane is tilted then only scalar multiplication of Eqs. 3.40 and 3.55 yields

$$\cos\theta_z = -\cos\delta \cos\alpha \cos\theta \cos\phi + \cos\delta \sin\alpha \sin\theta \cos\phi$$
$$- \sin\delta \cos\alpha \sin\theta - \sin\delta \sin\alpha \cos\theta \,. \qquad (3.59)$$

By considering the basic trigonometric relationships this expression can be rewritten succinctly as follows:

$$\cos\theta_z = -[\cos\delta \cos\phi \cos(\theta + \alpha) + \sin\delta \sin(\theta - \alpha)] \,, \qquad (3.60)$$

which reduces to Eq. 3.55 when $\alpha = 0$.

3.12 Solar Energy Calculations

Once the solar irradiance, I, on the ground is known then the solar radiation perpendicular to a horizontal surface I_H can be calculated similar to Eq. 3.11 as

$$I_H = I \cos\theta_z, \tag{3.61}$$

where θ_z is the zenith angle (see Fig. 3.20). Hourly direct radiation is obtained by integrating this quantity over a 1-h period:

$$I_h = \int_0^{1h} I \cos\theta_z \tag{3.62}$$

In the measurement of direct irradiation, I_d, two pyranometer readings are necessary, one with and the other without an occulting device. The hourly beam radiation on a horizontal surface is deduced from the difference between the these readings, thus,

$$I_H = I - I_d. \tag{3.63}$$

Hence, the diffuse radiation can be obtained as

$$I_d = I - I_H. \tag{3.64}$$

The daily global radiation, I_{Dg}, on a horizontal surface can be calculated by integration as

$$I_{Dg} = \int^{day} I \, dt. \tag{3.65}$$

Similarly, daily diffuse radiation, I_{Dd}, on a horizontal surface is

$$I_{Dd} = \int^{day} I_d \, dt. \tag{3.66}$$

Hence, daily direct radiation is the difference between these two quantities and can be written as

$$I_D = I_{Dg} - I_{Dd}. \tag{3.67}$$

It is possible to calculate the daily solar energy input incident on a sloping terrain. Considerations from Fig. 3.20 lead to the following expression for the daily solar radiation energy input, I_{Day}, onto a sloping surface as

$$I_{Day} = \int_{\phi_{sr}}^{\phi_{ss}} I \cos\theta_z \, d\phi, \tag{3.68}$$

where ϕ_{sr} and ϕ_{ss} are any hour angles after the sunrise and before the sunset, respectively, at the point A (Fig. 3.25). These angles expressed in radians in the limits of integration depend on the latitude, θ, and the solar declination, δ, as is the case for a horizontal surface, and, in addition, on the slope (α) and aspect (Ω) of an insolated surface, in general.

On the other hand, the monthly average daily values of the extraterrestrial irradiation on a horizontal plane $\overline{H}_0$ and the maximum possible sunshine duration $\overline{S}_0$ are two important parameters that are frequently needed in solar energy applications (Chap. 4). The values of $\overline{H}_0$ have been tabulated by Duffie and Beckman (1980) and Iqbal (1986) for latitude intervals of 5°. Most of the solar radiation researchers make their own calculations for these parameters. For reducing the amount of calculations, short-cut methods of using the middle day of each month or a single recommended day for each month have often been employed (Klein 1977). The values of H_0 for a given day can be computed by

$$H_0 = \frac{24 \times 3600}{\pi} I_0 \times \varepsilon \left(\cos\phi \cos\delta \sin\phi_{ss} + \frac{2\pi \phi_{ss}}{360} \sin\phi \sin\delta \right), \tag{3.69}$$

where ϕ is the latitude, δ is the declination angle, ϕ_{ss} is the sunset hour angle, and ε is the eccentricity given in Eqs. 3.23 or 3.26. The monthly averages of H_0 for each month can be calculated for the ϕ values from 90°N to 90°S at 1° intervals from Eq. 3.69.

On the other hand, values of monthly average daily maximum possible sunshine duration S_0 for a given day and latitude can be obtained from

$$S_0 = \frac{2}{15} \cos^{-1}(-\tan\phi \tan\delta). \tag{3.70}$$

The monthly averages, S'_0, can be taken for all the ϕ values. The values of S'_0 can be computed from the following expression:

$$S'_0 = \frac{2}{15} \cos^{-1} \left(\frac{\cos 85^0 - \sin\phi \sin\delta}{\cos\phi \cos\delta} \right). \tag{3.71}$$

The use of Eq. 3.64 instead of Eq. 3.63 can lead to substantial differences in the values of S_0 of up to 10%. Different solar radiation calculations can be obtained from already prepared tables given by ASHRAE (1981). A suitable site for such information is presented in http://www.solarviews.com.

3.12.1 Daily Solar Energy on a Horizontal Surface

Similar to Eq. 3.68, daily insolation, I_{DH}, and solar energy input due to direct radiation on a horizontal surface are given as

$$I_{DH} = \int_{\phi_{sr0}}^{\phi_{ss0}} \tau \varepsilon I_0 \cos\theta_z \, d\phi, \tag{3.72}$$

3.12 Solar Energy Calculations

where I_0 is the solar constant, τ is the total atmospheric transmissivity which assumes values between 0 and 1 and it depends on the hour angle (see Sect. 3.11), and ε is the eccentricity ratio (Eqs. 3.23 or 3.26). Hence, the radiation flux is equal to the fraction of solar constant as $\tau \varepsilon I_0$, which is equivalent to the terrestrial solar radiation as

$$I = \varepsilon \tau I_0 . \tag{3.73}$$

Additionally, in Eq. 3.72 θ_z is the zenith angle at any point with latitude, θ, and longitude, ϕ (see Fig. 3.34); and finally ϕ_{sr0} and ϕ_{ss0} are the sunrise and sunset hour angles, which are expressed in radians.

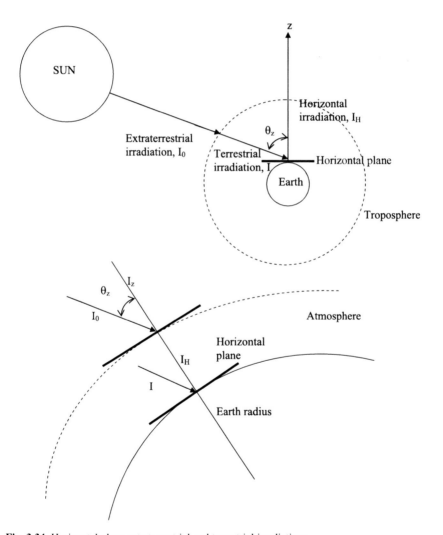

Fig. 3.34 Horizontal plane extraterrestrial and terrestrial irradiations

Substitution of the relevant quantities from Eq. 3.58 into Eq. 3.61 gives the normal incidental solar radiation on a horizontal surface in its most explicit form as

$$I_H = I_0 \varepsilon (\cos\delta \cos\theta \cos\phi + \sin\delta \sin\theta) . \tag{3.74}$$

This is the amount of solar radiation from the sun at any instant, but if the radiation amount during a specific time interval, say dt, is requested then due to the linear relationship with time the amount of solar energy becomes

$$dI_H = I_0 \varepsilon (\cos\delta \cos\theta \cos\phi + \sin\delta \sin\theta) dt . \tag{3.75}$$

If dt is in hours then the unit of this amount will be kJ/m^2/h. It is necessary to convert the time in hours to the hour angle by considering the rotational speed, ω, of the earth around it axis which leads to first $\omega = 2\pi/24 = d\phi/dt$ and, hence,

$$dt = \frac{12}{\pi} d\phi . \tag{3.76}$$

Its substitution into Eq. 3.63 leads to

$$I_H = \frac{12}{\pi} I_0 \varepsilon \tau (\cos\delta \cos\theta \cos\phi + \sin\delta \sin\theta) d\phi . \tag{3.77}$$

Practical applications require adaptation of a unit time interval as 1 h and in this case the i-th hour from the solar noon with ϕ_i hour angle at the midpoint of this period, the radiation over a period of 1 h can be calculated as

$$I_H = \frac{12}{\pi} I_0 \varepsilon \tau \int_{\phi_i - \pi/24}^{\phi_i + \pi/24} (\cos\delta \cos\theta \cos\phi + \sin\delta \sin\theta) d\phi . \tag{3.78}$$

After the necessary algebraic calculations, it becomes approximately

$$I_H = \frac{12}{\pi} I_0 \varepsilon \tau (\cos\delta \cos\theta \cos\phi + \sin\delta \sin\theta_i) . \tag{3.79}$$

It is possible to calculate the daily solar radiation amount, I_D, from this last equation by considering the sunrise and sunset hours at a particular point. It is assumed that from sunrise to noon time there will be an increase in the hourly solar radiation and then onward from noon to sunset there will be a decrease in a similar pattern. Hence, the hourly solar radiation variation within the day is considered as symmetrical with respect to noon time. Such a consideration gives the following simple equations:

$$I_D = \int_{\phi_{sr}}^{\phi_{ss}} I_H dt = 2 \int_0^{\phi_{ss}} I_H dt . \tag{3.80}$$

3.12 Solar Energy Calculations

The substitution of Eq. 3.72 into this expression gives

$$I_D = \frac{24}{\pi} I_0 \varepsilon \tau \left[\frac{\pi}{180} \phi_{ss} (\sin \delta \sin \theta) + \cos \delta \cos \phi \cos \phi_{ss} \right], \quad (3.81)$$

where ϕ_{ss} indicates the sunset time in degrees.

3.12.2 Solar Energy on an Inclined Surface

If the plane surface is tilted toward the equator then its position can be represented schematically as in Fig. 3.35.

At the interface of the atmosphere and space the horizontal plane, which is perpendicular to the direction of the earth's radius helps to define the extraterrestrial solar radiation onto a horizontal and tilted surface (see Fig. 3.34). Incident extraterrestrial solar beam radiation onto an inclined surface can be calculated as

$$I_\alpha = I \cos \theta_0 = \varepsilon \tau I_0 \cos \theta_0 \quad (3.82)$$

and onto a horizontal surface as

$$I_h = I_\alpha \cos \alpha \quad (3.83)$$

or

$$I_h = I_\alpha \cos \theta_0 \cos \alpha . \quad (3.84)$$

In these expressions the symbol I means incident terrestrial solar radiation per unit time and unit area (W/m²). Within one year on any day, the value of I can be calculated according to Eq. 3.15 by considering the extraterrestrial irradiation solar constant, I_0 (W/m²), which is taken as $I_0 = 1360$ W/m² in many practical studies. If the extraterrestrial solar radiation between two solar times ($t_2 > t_1$) is given for

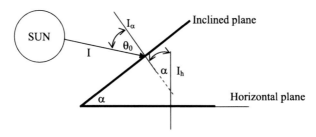

Fig. 3.35 Inclined plane terrestrial irradiation

one hour then these expressions can be integrated over solar time as

$$\int_{t_1}^{t_2} I_\alpha \, dt = I_{0\alpha} \tag{3.85}$$

and onto a horizontal surface as

$$\int_{t_1}^{t_2} I_z \, dt = I_{0z}. \tag{3.86}$$

Consideration of Eq. 3.61 in a similar manner to Eq. 3.70 and then its substitution into the last expression leads to:

$$I_\alpha = \frac{12}{\pi} I_0 \varepsilon \tau \int_{\phi_1}^{\phi_2} [\sin\delta \sin(\theta - \alpha) + \cos\delta \cos(\theta - \alpha)\cos\phi] \, d\phi. \tag{3.87}$$

Again for a unit time period of 1 h and with ϕ_M, which is the hour angle at mid-hour, it is possible to obtain from the last expression after integration, approximately,

$$I_\alpha = \frac{12}{\pi} I_0 \varepsilon \tau [\sin\delta \sin(\theta - \alpha) + \cos\delta \cos(\theta - \alpha)\cos\phi_M]. \tag{3.88}$$

If the radiation is required for a shorter duration than one hour, Δt, starting from t_1 and ending at t_2 ($\Delta t = t_1 - t_2 < 1$), then after the necessary calculations one can obtain

$$I_\alpha |_{t_1}^{t} = I_0 \varepsilon \left\{ \sin\delta \sin(\theta - \alpha)\Delta t + \frac{12}{\pi} \cos\delta \cos(\theta - \alpha) \left[\sin(15 t_1) - \sin(15 t_2) \right] \right\}. \tag{3.89}$$

Herein, t_1 and t_2 are in hours from midnight. It is possible to determine t on a particular day (that is, a particular declination, δ) that has irradiation equal to the monthly average hourly irradiation. Monthly average extraterrestrial hourly irradiation, $\overline{I}_I$, can be calculated as

$$\overline{I}_\alpha = I_\alpha |_{\delta = \delta_c}. \tag{3.90}$$

On the other hand, by considering the symmetry principle as in Eq. 3.73 the daily irradiation on a tilted plane after the integration of Eq. 3.69 from sunrise to sunset hour angles becomes

$$I_{\alpha \text{day}} = \frac{24}{\pi} I_0 \varepsilon \tau \int_0^{\phi = \min(\phi_z, \phi'_{\text{srt}})} [\sin\delta \sin(\theta - \alpha) + \cos\delta \cos(\theta - \alpha)\cos\phi] \, d\phi, \tag{3.91}$$

where ϕ_z is the solar hour corresponding to zenith time. It can be obtained by substituting $\theta_z = 0$ into Eq. 3.53, which gives

$$\phi_z = \cos^{-1}[\tan\delta \tan(\theta - \alpha)] . \tag{3.92}$$

If each of the sunset hour angles is considered as the minimum then the following two expressions can be obtained:

$$I_{\alpha\mathrm{day}} = \frac{24}{\pi} I_0 \varepsilon \tau \left[\frac{\pi}{180} \phi_{sr} \sin\delta \sin(\theta - \alpha) + \cos\delta \cos(\theta - \alpha) \sin\phi_{sr} \right], \tag{3.93}$$

(for $\phi_z \leq \phi_{sr}$),

and

$$I_{\alpha\mathrm{day}} = \frac{24}{\pi} I_0 \varepsilon \tau \left[\frac{\pi}{180} \phi_{sr} \sin\delta \sin(\theta - \alpha) + \cos\delta \cos(\theta - \alpha) \sin\phi_{sr} \right], \tag{3.94}$$

(for $\phi_{sr} \leq \phi_z$).

In these expressions ϕ_{sr} and ϕ_z should be substituted in degrees. Combination of these two equations leads to

$$I_{\alpha\mathrm{day}} = \frac{24}{\pi} I_0 \varepsilon \tau \left[\frac{\pi}{180} \phi_{sr} \sin\delta \sin(\theta - \alpha) + \cos\delta \cos(\theta - \alpha) \sin\phi_{sr} \right], \tag{3.95}$$

where ϕ_{sr} is $\min(\phi_{sr}, \phi_z)$. Finally, the monthly average extraterrestrial daily irradiation $\bar{I}_M$ on a tilted surface toward the equator can be calculated similar to the previous cases as

$$\bar{I}_{\alpha\mathrm{day}} = I_{\alpha\mathrm{day}}\big|_{\delta=\delta_c} . \tag{3.96}$$

3.12.3 Sunrise and Sunset Hour Angles

On any given day under clear-sky circumstances sunrise (sunset) occurs when solar radiation shines for the first (last) time upon a surface. The difference between the times of sunset and sunrise gives the duration of daily irradiation, which has more involved calculations in the case of double sunrise (sunset) time for certain combinations of slope, aspect, latitude, and solar declination.

During sunrise (sunset) the angle between the solar beam and the normal of the plane is $\theta_z = 90°$ and hence $\cos\theta_z = \cos 90 = 0$, the substitution of which into Eq. 3.56 leads to the hour angle equation as

$$A\cos\phi + B\sin\phi + C = 0 \tag{3.97}$$

where A, B, and C depend on the slope, aspect, latitude, and solar declination as follows:

$$A = -\cos\delta\cos\alpha\cos\theta + \cos\delta\sin\alpha\cos\Omega\sin\theta , \tag{3.98}$$

$$B = -\cos\delta\sin\alpha\sin\Omega , \tag{3.99}$$

and

$$C = -\sin\delta\cos\alpha\sin\theta - \sin\delta\sin\alpha\cos\Omega\cos\theta . \tag{3.100}$$

The general solution of Eq. 3.97 will be presented at the end of this section for calculating the sunrise (sunset) hour. Two different solutions will fall into the interval from 0 to π ($-\pi$ to 0) corresponding to the sunset, ϕ_{ss}^* (sunrise, ϕ_{sr}^*) hour. However, for the time being the simplest solutions are from the substitution of $\alpha = \Omega = 0$ first into Eqs. 3.98–3.100 and then Eq. 3.97 gives

$$\cos\phi = -\frac{C}{A} = -\tan\delta\tan\theta . \tag{3.101}$$

For this case the sunrise-hour and sunset-hour angles are

$$\phi_{sr0} = -\cos^{-1}(-\tan\delta\tan\theta) - \pi \le \phi_{sr0} \le 0 \tag{3.102}$$

and

$$\phi_{ss0} = \cos^{-1}(-\tan\delta\tan\theta)\ 0 \le \phi_{ss0} \le \pi , \tag{3.103}$$

respectively. On the other hand, in terms of the earth's angular rotation velocity, the sunrise and sunset times can be expressed as

$$t_{sr0} = \frac{\phi_{sr0}}{\omega} \tag{3.104}$$

and

$$t_{ss0} = \frac{\phi_{ss0}}{\omega} , \tag{3.105}$$

respectively. If $t_{sr0} < 0$ the sunrise precedes the solar noon. In general, the duration of daily insolation, D_{di}, is the difference between the sunrise and sunset times as given by

$$D_{di} = t_{sr0} - t_{ss0} = \frac{(\phi_{sr0} - \phi_{ss0})}{\omega} . \tag{3.106}$$

In the northern hemisphere, locations with a latitude $90° - \delta \le \theta \le 90°$ (with $0 > \delta$) are insolated 24 h daily, in which case, $\phi_{sr0} = -\pi$ and $\phi_{ss0} = \pi$.

By viewing the earth from over the north pole when $\delta = 23°45'$ (summer solstice), it is possible to determine the sunrise and sunset hour angles on a sloping surface (see Fig. 3.36).

3.12 Solar Energy Calculations

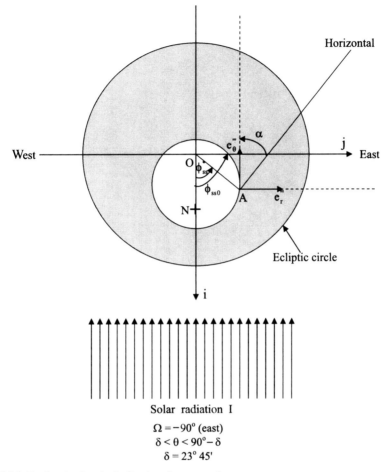

Fig. 3.36 North pole view for inclined surface sunset hour

In this figure a downward sloping plane is located at point A with aspect $\Omega = -90°$ due east. For a horizontal plane at point A the sunset hour angle, ϕ_{ss0}, is given by Eq. 3.103. Additionally, the theoretical sunset hour angle ϕ_{ss}^* from Eq. 3.97 is smaller than ϕ_{ss0}. Hence, the actual sunset hour angle is, $\phi_{ss} = \phi_{ss}^*$, because at angle ϕ_{ss0} the sloping plane is shaded by the curved shape of the earth.

On the other hand, Fig. 3.37 is the continuation of the situation with point A emerging from darkness.

The theoretical sunrise hour angle ϕ_{sr}^* from Eq. 3.97 is smaller with more negative values than ϕ_{ss0}. Hence, is does not equal the actual sunrise hour angle. However, the actual sunrise hour angle equals ϕ_{sr0}, the sunrise hour angle at A on a horizontal surface. This is due to the fact that at angle ϕ_{sr}^* the point A is under the earth's shadow geometry. The last two figures imply that the solutions from Eq. 3.103 for ϕ_{sr}^* and ϕ_{ss}^* equal the actual sunrise and sunset hour angles, respectively, when the

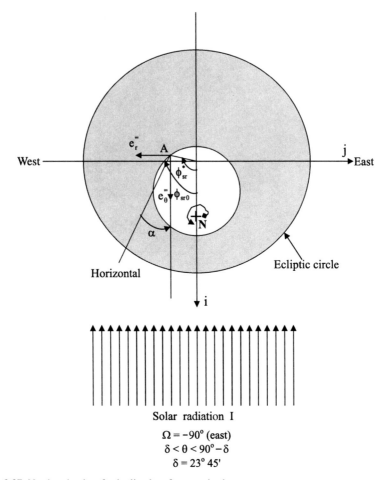

Fig. 3.37 North pole view for inclined surface sunrise hour

sloping plane is not shaded at ϕ_{sr}^* or at ϕ_{ss}^*. This same conclusion is valid in the most general plane position with any slope, α, and aspect, Ω angles.

Hence, it is possible to arrive at the following rules for determining the actual sunrise and sunset hour angles as

$$\phi_{sr} = \max(\phi_{sr}^*, \phi_{sr0}) \qquad -\pi \leq \phi_{sr} \leq 0 \qquad (3.107)$$

and

$$\phi_{ss} = \min(\phi_{ss}^*, \phi_{ss0}) \qquad 0 \leq \phi_{sr} \leq \pi , \qquad (3.108)$$

respectively. It has already been explained that some combinations of latitude, solar declination, slope, and aspect produce 24-h daily insolation, where $\phi_{sr} = -\pi$ and $\phi_{ss} = \pi$. After the determination of ϕ_{sr} and ϕ_{ss} they can then be substituted into Eq. 3.72 to calculate the daily energy input. Similar to Eq. 3.106 the duration of

3.12 Solar Energy Calculations

daily insolation can be obtained as

$$D_{di} = t_{ss} - t_{sr} = \frac{1}{\omega}(\phi_{ss} - \phi_{sr}), \qquad (3.109)$$

where t_{sr} and t_{ss} are the times of sunrise and sunset, respectively.

In order to solve Eq. 3.97 in a general form it is necessary to relate $\sin\phi$ and $\cos\phi$ terms through the well-known trigonometric relationship, $\sin^2\phi + \cos^2\phi = 1$, the substitution of which into Eq. 3.97 yields, after some algebra,

$$\left(A^2 + B^2\right)\cos^2\phi + 2AC\cos\phi + \left(C^2 - B^2\right) = 0. \qquad (3.110)$$

This expression has two roots, namely, for sunrise, ϕ_{sr}, and sunset, ϕ_{ss}. Although for a horizontal plane they are symmetrical, for inclined planes they are asymmetrical. The general solution of Eq. 3.110 gives

$$\phi_{sr} = \cos^{-1}\left[\frac{-AC - \sqrt{4A^2C^2 - 4\left(A^2 + B^2\right)\left(C^2 - B^2\right)}}{2\left(A^2 + B^2\right)}\right] \qquad (3.111)$$

and

$$\phi_{ss} = \cos^{-1}\left[\frac{-AC + \sqrt{4A^2C^2 - 4\left(A^2 + B^2\right)\left(C^2 - B^2\right)}}{2\left(A^2 + B^2\right)}\right]. \qquad (3.112)$$

In the case of a horizontal plane $\alpha = \Omega = 0$ and from Eqs. 3.98–3.100 $A = -\cos\delta\cos\theta$, $B = 0$, and $C = -\sin\delta\sin\theta$ and their substitutions into these last two expressions yield Eqs. 3.102 and 3.103, respectively.

3.12.3.1 Double Sunrise and Sunset

The northern (southern) high latitudes with steep slope produce two sunrises, (sunsets) ϕ_{sr1}, ϕ_{sr2}, with $\phi_{sr1} > \phi_{sr2}$ (ϕ_{ss1}, ϕ_{ss2}, with $\phi_{ss2} > \phi_{ss1}$) hour angles. For instance, if the critical slope in Fig. 3.29 is exceeded then a shaded slope is produced during an interval that would otherwise be in light. This slope may be insolated prior to and after that interval of darkness hence causing to two sunrises and two sunsets. The first sunrise occurs when the sun first shines on the slope on any clear-sky day. The first sunset ends the first period of insolation, at which time darkness sets in on the slope until the second sunrise (ϕ_{sr2}) restarts insolation. The latter ends with the second sunset (ϕ_{ss2}). In this case the energy input Eq. 3.68 can be expressed as

$$I_S = \int_{\phi_{sr1}}^{\phi_{ss1}} I \cdot \mathbf{u}_r''' \, d\phi + \int_{\phi_{sr2}}^{\phi_{ss2}} I \cdot \mathbf{u}_r''' \, d\phi, \qquad (3.113)$$

where all the intervening terms are defined exactly in association with Eq. 3.59.

The angles are expressed in radians. Hence, $\phi_{sr1} = \phi_{sr0}$, $\phi_{ss1} = \phi_{sr}^*$, $\phi_{sr2} = \phi_{ss}^*$, and $\phi_{ss2} = \phi_{ss0}$, where ϕ_{sr0} and ϕ_{ss0} are defined in Eqs. 3.107 and 3.108, respectively, and correspond to the horizontal-case hour angles. In the case of double sunrise and sunset situations the duration of daily insolation can be expressed as follows:

$$D_{di} = \frac{1}{\omega}(\phi_{ss1} - \phi_{sr1} + \phi_{ss2} - \phi_{sr2}) \,. \tag{3.114}$$

References

ASHRAE (1981) Handbook of fundamentals, chapter 27. American Society of Heating Refrigerating and Air Conditioning Engineers, New York

Barbour MC, Burk JH, Pitts WD (1978) Terrestrial plant ecology, 2nd edn. Cummings, Menlo Park

Becquerel AE (1839) Recherges sur les effets de la radiation chimique de la lumiere solaire au moyen des courants electriques produits sous l'influence des rayons solaires. C R Acad Sci 9:145–149, 561–567

Cartwright TJ (1993) Modeling the world in a spreadsheet: environmental simulation on a microcomputer. Johns Hopkins University Press, Baltimore

Collares-Pereira M, Rabl A (1979) The average distribution of solar radiation correlations between diffuse and hemispherical and between daily and hourly insolation values. Solar Energy 22:155–164

Dubayah R (1992) Estimating net solar radiation using Landsat Thematic Mapper and digital elevation data. Water Resour Res 28:2469–2484

Dubayah R, Rich PM (1995) Topographic solar radiation models for GIS. Int J Geographical Information Systems. 9:405–419

Dubayah R, Dozier J, Davis F (1989) The distribution of clear-sky radiation over varying terrain. In: Proceedings of the international geographic and remote sensing symposium, European Space Agency, Neuilly, 2:885–888

Duffie JA, Beckman WA (1980) Solar engineering thermal processes. Wiley, New York

Duffie JA, Beckman WA (1991) Solar engineering of thermal processes. Wiley, New York

Dunn PD (1986) Renewable energies: sources, conversion and application. Peregrinus, Cambridge

Forster BC (1984) Derivation of atmospheric correction procedures for LANDSAT MSS with particular reference to urban data. Int J Remote Sensing 5:799–817

Frochlich C, Werhli C (1981) Spectral distribution of solar irradiation from 2500 to 250. World Radiation Centre, Davos

Frouin R, Lingner DW, Gautier C, Baker KS, Smith RC (1989) A simple analytical formula to compute clear sky total and photo-synthetically available solar irradiance at the ocean surface. J Geophysical Res 94:9731–9742

Gates DM (1980) Biophysical ecology. Springer, New York

Graves C (1998) Reflected radiation. http://quake.eas.slu.edu/People/CEGraves/Eas107/notes/node25.html. Accessed 5 August 2000

Hay JE (1984) An assessment of the meso-scale variability of solar radiation at the Earth's surface. Solar Energy 32:425–434

Hottel HC (1976) A simple model for estimating the transmittance of direct solar radiation through clear atmospheres. Solar Energy 18:129–134

Iqbal M (1986) An introduction to solar radiation. Academic, Toronto

Jain PC (1988) Accurate computations of monthly average daily extraterrestrial irradiation and the maximum possible sunshine duration. Solar Wind Technol 5:41–45

Klein SA (1977) Calculation of monthly average insolation on tilted surfaces. Solar Energy 19:325

References

Kondratyev KY (1965) Radiative heat exchange in the atmosphere. Pergamon, New York
Kreider JF, Kreith F (1981) Solar energy handbook. McGraw-Hill, New York
Kreith F, Kreider JF (1978) Principles of solar engineering. McGraw-Hill, New York
Lamm LO (1981) A new analytic expression for the equation of time. Solar Energy 26:465
Liu BY, Jordan RC (1960) The interrelationship and characteristic distribution of direct, diffuse and total solar radiation. Solar Energy 4:1–19
McAlester AL (1983) The earth: an introduction to the geological and geophysical sciences. Prentice-Hall, Englewood Cliffs
Monteith JL (1962) Attenuation of solar radiation: a climatological study. Q J Meteorol Soc 88:508–521
Monteith JL, Unsworth MH (1990) Principles of environmental physics. Arnold, London
Richards JA (1993) Remote sensing digital image analysis: an introduction. Springer, Berlin
Şen Z (2004) Solar energy in progress and future research trends. Progr Energy Combustion Sci Int Rev J 30:367–416
Spencer JW (1972) Fourier series representation of the position of the sun. Search 2:172
Stacey FD (1992) Physics of the earth. Brookfield, Brisbane
Williams LD, Barry RG, Andrews JT (1972) Application of computed global radiation for areas of high relief. J Appl Meteorol 11:526–533
Woolf HM (1968) Report NASA TM-X-1646. NASA, Moffet Field

Chapter 4
Linear Solar Energy Models

4.1 General

Long-term average values of the instantaneous (or hourly, daily, monthly) global and diffuse irradiation on a horizontal surface are needed in many applications of solar energy designs. The measured values of these parameters are available at a few places. At others no measurements exist and here the usual practice is to estimate them from theoretical or empirical models that have been developed on the basis of measured values.

In practical studies it is a logical and rational idea that the solar radiation is directly proportional to the sunshine duration. The formulation of the proportionality can be derived from the measurements of the variables through scatter diagrams and most often by the application of statistical regression methods. In this book, the reader is advised to look at the scatter diagrams visually and make interpretations prior to any modeling attempt. Like in any other discipline of science, early solar energy models have linear mathematical forms similar to scientific laws (Newton, Hooke, Fourier, Fick, Hubble, Ohm, Darcy), which express linear relationships between two relevant variables. For instance, provided that the mass is constant the force is directly and linearly proportional to the acceleration. Similarly, in the solar energy literature, before entering into more complicated and sophisticated models, the original models expressed the relationship between solar radiation and the sunshine duration as a straight line. Such a pioneering relationship was presented by Angström in 1924. The first attempt to analyze the hourly radiation data is due to Hoyt (1978) who employed the data of widely separated localities to obtain the curves or the ratio (hourly/daily) for the observed global radiation versus the sunset hour angle for each hour from 9 a.m. to 3 p.m. Liu and Jordan (1960) extended the day length of these curves.

Knowledge of the amount of solar radiation falling on a surface of the earth is of prime importance to engineers and scientists involved in the design of solar energy systems. In particular, many design methods for thermal and photovoltaic systems require monthly average daily radiation on a horizontal surface as an input, in order to predict the energy production of the system on a monthly basis (Beckman et al., 1977; Ma and Iqbal 1984; Thevenard et al., 2000).

This chapter will first provide the fundamental assumptions in a linear model, such as the Angström model and then several alternatives with the exclusion of a few restrictive assumptions. The whole modeling procedure revolves around the plausible estimation of the model parameters from a given set of data. In general, Angström model parameter estimations are achieved through the application of a classic statistical regression approach, which has a set of restrictive assumptions that are not taken into account in almost all the practical applications.

4.2 Solar Radiation and Daylight Measurement

Systematic measurements of *diffuse* solar energy and the *global* (total) irradiation incident on a horizontal surface are usually undertaken by a national agency, which is the national meteorological office in many countries. The measurement network includes *pyranometers*, *solarimeters*, or *actinography* instruments for this purpose. At several locations direct or beam irradiation is measured by a *pyrheliometer* with a fast-response multi-junction thermopile. Diffuse irradiance is measured at a set of stations by placing a shadow band over a pyranometer. In practice, it is very important to appreciate the order of measurements prior to any modeling study both for *solar radiation* and *sunshine duration* or daylight. The present state of solar radiation and daylight models is such that they are approaching the accuracy limits set out by the measuring equipment (Gueymard 2003; Perez et al., 1990). Radiation in the visible region of the spectrum is often evaluated with respect to its visual sensation effect on the human eye.

There is a relative abundance of sunshine duration data and therefore it is a common practice to correlate the solar radiation to sunshine duration measurements. In many countries, diurnal bright sunshine duration is measured at a wide number of places. The hours of bright sunshine are the time during which the sun's disk is visible. It has been measured using the well-known Campbell–Stokes *sunshine recorders*, which use a solid glass spherical lens to burn a trace of the sun on a graduated paper. It produces the trace whenever the beam irradiation is above a critical level. Although the critical threshold varies loosely with the prevailing ambient conditions, the sunshine recorder is an economic and robust device and hence it is used widely. The limitations of the Campbell–Stokes sunshine recorder are well known and have been discussed in the Observers' Handbook (1969), Painter (1981), and Rawlins (1984). Some of the associated limitations with this device are that the recorder does not register a burn on the card below a certain level of incident radiation, which is about $150-300 \, W/m^2$. On a clear day with a cloudless sky the burn does not start until $15-30$ min after sunrise and usually ceases about the same period before sunset. This period varies with the season. On the other hand, under periods of intermittent bright sunshine the burn spreads. The diameter of the sun's

4.2 Solar Radiation and Daylight Measurement

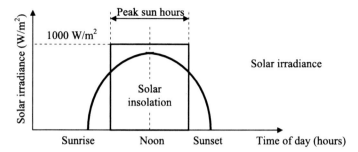

Fig. 4.1 Peak sun hours

image formed by the spherical lens is only about 0.7 mm. However, a few seconds of exposure to bright sunshine may produce a burn width of about 2 mm. As such, intermittent sunshine may be indistinguishable from a longer period of continuous sunshine.

The results of the earth's motion and the atmospheric effects at various locations have led to essentially two types of solar insolation data as the average daily and hourly data. Unlike irradiation, which is defined as the solar power per unit area, solar radiation is radiant energy per unit area. Solar radiation is determined by summing solar irradiance over time and it is expressed usually in units of kW/m^2 per day.

The number of peak sun hours per day at a given location is the equivalent time (in hours) at peak sun condition ($1000\,W/m^2$) that yields the same total insolation. Figure 4.1 shows how peak sun hours are determined by constructing a graph having the same area as that for the actual irradiation versus time.

In order to homogenize the data of the worldwide network for sunshine duration, a special design of the Campbell–Stokes sunshine recorder, the so-called Interim Reference Sunshine Recorder (IRSR), was recommended as the reference (WMO 1962). According to WMO (2003), sunshine duration during a given period is defined as the sum of that sub-period for which the direct solar irradiance exceeds $120\,W/m^2$.

4.2.1 Instrument Error and Uncertainty

Any measurement includes systematic, random, and equipment inherent errors. Angus (1995) has provided an account of the measurement errors associated with solar radiation and sunshine duration measurements. He stated that the most common error sources arise from the sensors and their construction. Among such error types are the following:

1. Cosine response
2. Azimuth response
3. Temperature response

4. Spectral selectivity
5. Stability
6. Non-linearity
7. Thermal instability
8. Zero offset due to nocturnal radiative cooling

Pyranometers in use have to meet the specifications set out by World Meteorological Organization (WMO). The *cosine effect* error is the most apparent and widely recognized error, which is the sensor's response to the angle at which radiation strikes the sensing area. The more acute the angle of the sun (at sunrise and sunset), the greater this error at altitude angles of the sun below 6°. This error source can be avoided through the exclusion of the recorded data at sunrise and sunset times.

The *azimuth angle error* appears as a result of imperfections of the glass domes. This is an inherent manufacturing error which yields a similar percentage error to the cosine effect. Like the azimuth error, the temperature response of the sensor is an individual fault for each cell. The photometers are thermostatically controlled and hence the percentage error due to fluctuations in the sensor's temperature is reduced. However, some pyranometers have a much less elaborate temperature control system. The pyranometers rely on the two glass domes to prevent large temperature swings. Ventilation of the instrument is an additional recommended option.

The spectral selectivity of some pyranometers is dependent on the spectral absorbance of the black paint and the spectral transmission of the glass. The overall effect contributes only a small percentage error to the overall measurements. Each sensor possesses a high level of stability with the deterioration of the cells resulting in approximately $\pm 1\%$ change in the full-scale measurement per year. Finally, the non-linearity of the sensors is a concern especially with photometers. It is a function of illuminance or irradiance levels and tends to contribute only a small percentage error toward the measured values. Table 4.1 provides details of the above-mentioned uncertainties.

In addition to the above sources of equipment-related errors care must be taken to avoid operational errors such as incorrect sensor leveling and orientation of the vertical sensors as well as improper screening of the vertical sensors from ground-reflected radiation.

4.2.2 Operational Errors

The sources of operation-related errors are self-explanatory and they can be categorized as follows:

1. Complete or partial shade-ring misalignment
2. Dust, snow, dew, water droplets, bird droppings, *etc.*
3. Incorrect sensor leveling
4. Shading caused by building structures

4.2 Solar Radiation and Daylight Measurement

5. Electric fields in the vicinity of cables
6. Mechanical loading on cables
7. Orientation and/or improper screening of the vertical sensors from ground-reflected radiation
8. Station shut down
9. Improper application of diffuse shade-ring correction factor
10. Inaccurate programming of calibration constants

It is good practice to protect cables from strong electric fields such as elevator shafts. Another source of error that may arise is from cables under mechanical load (piezoelectric effects), which is the production of electrical polarization in a material by the application of mechanical stress. Failure to protect cables from the above sources may produce "spikes" in the data and these are shown as unusually high values of solar irradiance. Figure 4.2 demonstrates the sources of error categorized under items (1) and (2) discussed above.

Such errors are best highlighted via a scatter diagram by plotting the *diffuse ratio* (the ratio of horizontal sky diffuse and the total or global irradiance) against *clearness index* (the ratio of horizontal global to extraterrestrial irradiance). Any serious departure of data from the normally expected trend can hence be identified.

4.2.3 Diffuse-Irradiance Data Measurement Errors

Historically, meteorological offices worldwide have used a *shade-ring correction* procedure that is based on the assumption of an *isotropic sky*. However, during the past 15 years a number of alternate, more precise methods that are based on a realistic, anisotropic sky have been established (Kreider and Kreith 1981; Perez et al., 1990). Old isotropic-sky-corrected diffuse-irradiation records are slightly high

Table 4.1 WMO classification of pyranometers

Characteristics	Secondary standard	First class	Second class
Resolution (smallest detectable change in W/m^2)	±1	±5	±10
Stability (percentage of full scale, change per year)	±1	±2	±5
Cosine response (% deviation from ideal at 10° solar elevation on a clear day)	±3	±7	±15
Azimuth response (% deviation from ideal at 10° solar elevation on a clear day)	±3	±5	±10
Temperature response (% maximum error due to change of ambient temperature within the operating range)	±1	±2	±5
Non-linearity (% of full scale)	±0.5	±2	±5
Spectral sensitivity (% deviation from mean absorbance 0.3 – 3 µm)	±2	±5	±10
Response time (99% response)	<25 s	<60 s	<240 s

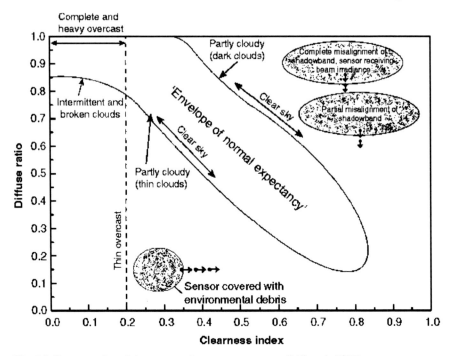

Fig. 4.2 Demonstration of the sources of measurement errors (Löf *et al.*, 1966)

(low) for overcast conditions by up to 10% for clear-sky conditions. It is imperative that due care is taken in using a precise and validated shade-ring correction procedure since any errors in horizontal diffuse-irradiance records will be multiplied by a large factor when horizontal beam irradiance and subsequently the total slope energy computations are undertaken.

Drummond (1965) estimates that accuracies of 2–3% are attainable for daily summations of radiation for pyranometers of first-class classification. Individual hourly summations even with carefully calibrated equipment may be in excess of 5%. Coulson (1975) infers that the errors associated with routine observations may be well in excess of 10%. There are isolated cases of poorly maintained equipment, but those that are in the regular network may exhibit monthly averaged errors of about 10%. However, not all designs of the latter sensor can claim even this level of accuracy. These figures must be borne in mind when evaluating the accuracy of the relevant models.

4.3 Statistical Evaluation of Models

The accuracy of the solar radiation mathematical models is important not only in the final stages of projects, but particularly in the initial stages prior to any systematic

4.3 Statistical Evaluation of Models

model construction work. Apart from the scatter diagram inspections, examination of residuals (model errors) is also recommended after the model establishment. The procedure is to produce a graph of the model error, e, which is the difference between measurements, Y_i and corresponding model estimates, $\hat{Y}_i (i = 1, 2, \ldots, n)$. These errors, $(e_i = Y_i - \hat{Y}_i)$, are plotted against the independent variable X_i on a Cartesian coordinate system (Fig. 4.3). In such a graph there may arise various alternative patterns which are:

1. Adequate model: If the scatter of points appears as in Fig. 4.3a then the model is adequate, because the model errors are scattered independently from each other.
2. Transformation model: If the band of error scatter widens (or narrows) as X_i or Y_i increases, as in Fig. 4.3b, then it indicates a lack of constant variance of the residuals, which is one of the violations of the regression model validity. The corrective measure in this case is a transformation of the Y variable until the error scatter appears as in Fig. 4.3a.
3. Linear independent model: A plot of the residuals such as in Fig. 4.3c indicates the absence of an independent variable in the model under examination, which is not a suitable model.
4. Non-linear independent model: If the scatter of points is as in Fig. 4.3d then a non-linear term must be added to the initial model.

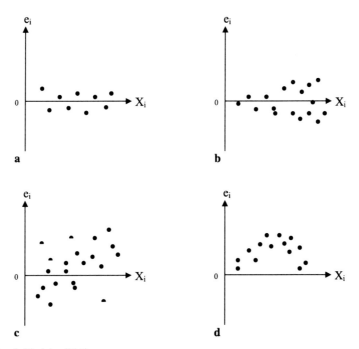

Fig. 4.3 a–d. Model validation

Another significant verification approach in modeling is the scatter diagram of measurements against corresponding model estimates. In the case of an adequate model the scatter of points should be close to the 45° line in a random manner as in Fig. 4.4 (Şen 2001a).

In practical studies, especially in the modeling of solar radiation from sunshine duration measurements, it is not possible to obtain a perfect plot on a 45° straight line. Less than ±10% deviations from the 45° line are acceptable in practical studies (Şen 2001a). However, some researchers may prefer ±5% or less. Figure 4.5 presents some cases that can be encountered in modeling, but they are not exhaustive and there may be other scatter patterns, which must be interpreted and the model adjustment made accordingly.

In the next sections of this chapter and in Chaps. 5 and 6 a number of models will be presented wherein one dependent variable is regressed against one or several independent variables or additional non-linear terms.

Often *correlation* between two quantities is also to be examined. In solar energy literature it has become common practice to refer to regression models as "correlation equations" based on the well-known least squares method. Correlation is

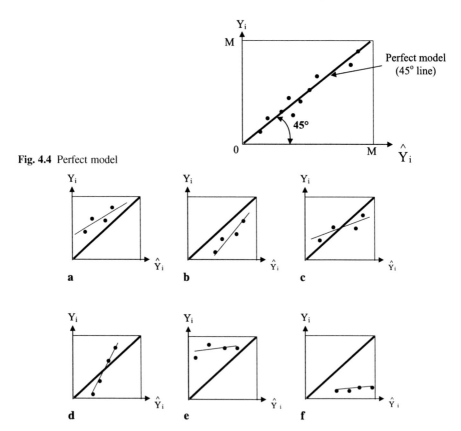

Fig. 4.4 Perfect model

Fig. 4.5 a–f. Scatter diagrams of measurement-model values

the degree of relationship between variables, which helps to seek determination of how well a linear model describes the relationship. On the other hand, regression is a technique of fitting linear or non-linear models between a set of n dependent, Y_i, and independent $X_i (i = 1, 2, \ldots, n)$ variables. In solar energy modeling most often a simple regression equation is used with one independent variable, X, in the form with constants a and b as

$$Y = a + bX . \qquad (4.1)$$

Provided that n pairs of measurements (X_i, Y_i) are available then the model parameter estimations can be found from the simple classic regression approach as

$$a = \overline{Y} - b\overline{X} \qquad (4.2)$$

and

$$b = \frac{\overline{YX} - \overline{X}\,\overline{Y}}{\overline{X^2} - \overline{X}^2} , \qquad (4.3)$$

where over-bars indicate the arithmetic averages of the attached variable (Davis 1986). For instance, $\overline{X}$ is the arithmetic average of X. On the other hand, there are also non-linear models between the solar radiation and sunshine duration as will be explained in Chap. 5. However, one alternative of such models may have the following quadratic and power mathematical forms:

$$Y = a + bX + cX^2 \qquad (4.4)$$

or

$$Y = a + bX^c , \qquad (4.5)$$

where a, b, and c are model parameters.

4.3.1 Coefficient of Determination (R^2)

In statistics literature, it is the proportion of variability in a data set that is accounted for by a statistical model, where the variability is measured quantitatively as the sum of square deviations. Most often it is defined notationally as

$$R^2 = \frac{\sum_{i=1}^{n} \left(\hat{Y}_i - \overline{Y}\right)^2}{\sum_{i=1}^{n} \left(Y_i - \overline{Y}\right)^2} . \qquad (4.6)$$

This can also be expressed as

$$R^2 = 1 - \frac{\sum_{i=1}^{n}\left(Y_i - \hat{Y}_i\right)^2}{\sum_{i=1}^{n}\left(Y_i - \overline{Y}\right)^2} \quad \left(0 \le R^2 \le 1\right). \tag{4.7}$$

Herein, Y_i and $\hat{Y}_i$ are the measurements and model estimates, respectively. A high value of R^2 is desirable as this shows a lower unexplained variation. R^2 is a statistic that gives some information about the goodness-of-fit of a model. In regression, the R^2 coefficient of determination is a statistical measure of how well the regression line approximates the real data points. An R^2 of 1.0 indicates that the regression line perfectly fits the data, which is never valid in any solar radiation estimation model.

4.3.2 Coefficient of Correlation (r)

The correlation coefficient implies the strength and direction of a linear relationship between two variables. In general, its statistical usage refers to the departure of two variables from independence. In this broad sense, there are several coefficients, measuring the degree of correlation. For instance, the square root of the coefficient of determination is defined as the coefficient of correlation, $-1 < r = \sqrt{R^2} < 1$. It is a measure of the relationship between variables based on a scale ranging between -1 and $+1$. Whether r is positive or negative depends on the inter-relationship between X_i and Y_i, i.e., whether they are directly proportional (high Y_i values follow high X_i values) or *vice versa*. Once r has been estimated for any fitted model, its numerical value may be interpreted as follows. For instance, if for a given regression model $r = 0.9$, it means that $R^2 = 0.81$. It may be concluded that 81% of the variation in Y has been explained (removed) by the model under discussion, leaving 19% to be explained by other factors.

The significance of r can be checked by *Student's t-test statistic*, which is given as

$$t = \sqrt{n-2}\left(\frac{r}{\sqrt{1-r^2}}\right), \tag{4.8}$$

where n is the number of data points and $(n-2)$ is the degrees of freedom (d.f.). If for a given location a regression model between average clearness index (K_T) and monthly averaged sunshine fraction (S/S_0) gives $r^2 = 0.64$ for 12 pairs of data points, then $t = (12-2)^{0.5}\left\{0.8/\sqrt{(1-0.64)}\right\} = 4.216$. In this example there are 10 d.f.. Thus from Table 4.2 the value of $r = 0.8$ is significant at 99.8% but not at 99.9% (note that for d.f. = 10, $t = 4.216$ lies between 4.144 and 4.587 corresponding to columns for 0.998 and 0.999, respectively). In general terms, this means that the regression model may yield estimates with 99.8% confidence.

4.3 Statistical Evaluation of Models

Table 4.2 Percentile values for Student's t-distribution

d.f.	Significance level				
	0.95	0.98	0.99	0.998	0.999
	12.706	31.821	63.657	318.310	636.620
2	4.303	6.965	9.925	22.327	31.98
3	3.182	4.541	5.841	10.214	12.924
4	2.76	3.747	4.604	7.173	8.610
5	2.571	3.365	4.032	5.893	6.869
6	2.447	3.143	3.707	5.208	5.959
7	2.365	2.998	3.499	4.785	5.408
8	2.306	2.896	3.355	4.501	5.041
9	2.262	2.821	3.250	4.297	4.781
10	2.228	2.764	3.169	4.144	4.587
15	2.131	2.602	2.947	3.733	4.073
20	2.086	2.528	2.845	3.552	3.850
25	2.060	2.485	2.787	3.450	3.725
30	2.042	2.457	2.750	3.385	3.646
40	2.021	2.423	2.704	3.307	3.551
60	2.000	2.390	2.660	3.232	3.460
120	1.980	2.358	2.617	3.160	3.373
200	1.972	2.345	2.601	3.131	3.340
500	1.965	2.334	2.586	3.107	3.310
1000	1.962	2.330	2.581	3.098	3.300
∞	1.960	2.326	2.576	3.090	3.291

4.3.3 Mean Bias Error, Mean of Absolute Deviations, and Root Mean Square Error

In order to gain further insight into the performance evaluation of a model, *mean bias error* (MBE), *mean absolute deviation* (MAD), and *root mean square error* (RMSE) may be defined in sequence as follows:

$$\text{MBE} = \frac{1}{n} \sum_{i=1}^{n} \left(Y_i - \hat{Y}_i \right), \tag{4.9}$$

$$\text{MAD} = \frac{1}{n} \sum_{i=1}^{n} \left| Y_i - \hat{Y}_i \right|, \tag{4.10}$$

and

$$\text{RMSE} = \left[\frac{1}{n} \sum_{i=1}^{n} \left(Y_i - \hat{Y}_i \right)^2 \right]^{1/2}. \tag{4.11}$$

The MBE is given as the arithmetic average of the errors. If its value is equal to zero, it does not mean that the model yields estimations without error. The MBE provides a measure of the overall trend of a given model, *i. e.*, predominantly over-estimating (positive values) or under-estimating (negative values). However, the smaller the MBE the better is the model result.

On the other hand, in an acceptable model, the MAD value should be as close as possible to zero, but never equal to zero in the solar radiation modeling.

The RMSE is similar to the MAD and provides a measure of squared deviations. In statistics, the RMSE of an estimator is the square root of the expected value of the square of the "error." The error is the amount by which the model estimate differs from the corresponding measurement. The error occurs because of randomness or the model does not account for information that could produce a more accurate estimate.

These error formulations provide quantitative measures which have the same physical units as the dependent variable, Y_i. In some instances, non-dimensional versions of MBE (NDMBE), MAD (NDMAD), and RMSE (NDRMSE) are required, which are defined simply as follows:

$$\text{NDMBE} = \frac{1}{n} \sum_{i=1}^{n} \left(\frac{Y_i - \hat{Y}_i}{\hat{Y}_i} \right), \tag{4.12}$$

$$\text{NDMAD} = \frac{1}{n} \sum_{i=1}^{n} \left| Y_i - \hat{Y}_i \right|, \tag{4.13}$$

and

$$\text{NDRMSE} = \frac{1}{n} \left[\left(\frac{Y_i - \hat{Y}_i}{\hat{Y}_i} \right)^2 \right]^{1/2}. \tag{4.14}$$

4.3.4 Outlier Analysis

Often in solar radiation studies one encounters data that lie unusually far removed from the bulk of the data population. Such data are called "*outliers*." One definition of an outlier is that it lies three or four standard deviations or more from the mean of the data population. The outlier indicates peculiarity and suggests that the datum is not typical of the rest of the data. As a rule, an outlier should be subjected to particularly careful examination to see whether any logical explanation may be provided for its peculiar behavior.

Automatic rejection of outliers is not always very wise. Sometimes an outlier may provide information that arises from unusual conditions. Outliers may however be rejected if the associated errors may be traced to erroneous observations due to any one or a combination of factors. Statistically, a "*near outlier*" is an observation that lies outside 1.5 times the inter-quartile range, which is the interval from the 1st

quartile to the 3rd quartile. The near outlier limits are mathematically defined as follows:
Lower outlier limit: 1st quartile $-$ 1.5 (3rd quartile $-$ 1st quartile)
Upper outlier limit: 3rd quartile $+$ 1.5 (3rd quartile $-$ 1st quartile)

Likewise, far outliers are defined as the data whose limits are defined below:
Lower limit: 1st quartile $-$ 3 (3rd quartile $-$ 1st quartile)
Upper limit: 3rd quartile $+$ 3 (3rd quartile $-$ 1st quartile)

4.4 Linear Model

The most widely used and the simplest equation relating radiation to sunshine duration is the Angström-Prescott relationship (Angström 1924; Prescott 1940), which can be expressed as a linear regression expression

$$\frac{\overline{H}}{\overline{H}_0} = a + b\frac{n}{N}, \qquad (4.15)$$

where $\overline{H}$ is the *monthly average daily radiation* on a horizontal surface, $\overline{H}_0$ is the monthly average daily *horizontal extraterrestrial radiation*, n is the number of *hours of bright sunshine per month*, N is the total number of daylight hours in the month, and, finally, a and b are model constants that should be determined empirically from a given data set. Angström (1929), Gueymard *et al.* (1995), Şahin *et al.* (2001), and Wahab (1993) assume a wide range of values depending on the location considered. If it is not possible to estimate these parameters from measured data for a specific location, they can be inferred from correlations established at neighboring locations as will be explained in Chap. 6 (Palz and Greif 1996; Şen and Şahin 2001).

The empirical determination of a and b is the greatest shortcoming of the Angström-Prescott relationship and it limits the usefulness of the formula. The Suehrcke derivation (Suehrcke 2000; Suehrcke and McCormick 1992) is presented here briefly. For a given month with a number of hours n of bright sunshine, the clear air sunshine fraction f_{clear} is defined as

$$f_{\text{clear}} = \frac{n}{N}, \qquad (4.16)$$

where N is the total number of daylight hours in the month. Suehrcke (2000) equates this approximately to

$$\frac{\overline{H}_{\text{b}}}{\overline{H}_{\text{b,clean}}}, \qquad (4.17)$$

where $\overline{H}_{\text{b}}$ is the monthly average of daily horizontal surface beam (direct) radiation and $\overline{H}_{\text{b,clean}}$ is the monthly average of daily clear-sky horizontal surface beam radiation. In order to relate $\overline{H}_b$ to monthly mean daily horizontal surface radiation $\overline{H}$, Suehrcke uses the Page (1961) diffuse fraction relationship as

$$\frac{\overline{H}_d}{\overline{H}} = 1 - C\overline{K}, \tag{4.18}$$

where $\overline{H}_d$ is the monthly mean daily horizontal surface diffuse radiation, C is a constant, and $\overline{K}$ is the monthly mean daily *clearness index* defined as

$$\overline{K} = \frac{\overline{H}}{\overline{H}_0}, \tag{4.19}$$

with $\overline{H}_0$ the monthly mean daily horizontal extraterrestrial radiation. Given that by definition

$$\overline{H} = \overline{H}_b + \overline{H}_d \tag{4.20}$$

and considerations from Eqs. 4.18–4.20 lead to

$$\overline{H}_b = C\overline{H}_0\overline{K}^2, \tag{4.21}$$

the same relationship for $\overline{H}_{b,\text{clean}}$ is

$$\overline{H}_{b,\text{clear}} = C\overline{H}_0\overline{K}^2_{\text{clear}}, \tag{4.22}$$

where $\overline{K}_{\text{clear}}$ is the monthly average clear-sky clearness index defined as

$$\overline{K}_{\text{clear}} = \frac{\overline{H}_{\text{clear}}}{\overline{H}_0}, \tag{4.23}$$

where $\overline{H}_{\text{clear}}$ is the monthly mean daily horizontal surface clear-sky radiation. Elimination of the constant C leads to Suehrcke's relationship

$$\overline{f}_{\text{clear}} = \left(\frac{\overline{K}}{\overline{K}_{\text{clear}}}\right)^2. \tag{4.24}$$

The only semi-empirical constant is $\overline{K}_{\text{clear}}$, which is a measurable quantity and it depends on the local atmospheric conditions and according to Suehrcke (2000) it is typically between 0.65 and 0.75.

On the other hand, by definition bright sunshine duration s is the number of hours per day that the sunshine intensity exceeds some predetermined threshold of brightness. Angström (1924, 1929) proposed a linear relationship between the ratio of monthly averaged global radiation $\overline{H}$ to cloudless global irradiation H_{cg} and monthly averaged sunshine duration, $\overline{s}$ leading to

$$\frac{\overline{H}}{H_{cg}} = c_1 + (1 - c_1)\frac{\overline{s}}{\overline{S}}, \tag{4.25}$$

4.4 Linear Model

where $c_1 = 0.25$ and $\overline{S}$ is the monthly averaged astronomical day duration (*day length*). Angström (1929) determined the value of c_1 from Stockholm data, but it was not until more than 30 years later that he (Angström 1956) stated that Eq. 4.25 was obtained from mean monthly data and should not be used with daily data.

In order to eliminate $\overline{H}$ from sunshine records, Angström's model required measurements of global radiation on completely clear days, H_{cg}. The limitation prompted Prescott (1940) to develop a model that was a fraction of the extraterrestrial radiation on a horizontal surface $\overline{H}_0$ rather than H_{cg}, since $\overline{H}_0$ can be easily calculated. Hence, the modified Angström model, referred to as the Angström-Prescott formula (Gueymard et al., 1995; Martinez-Lozano et al., 1984) is

$$\frac{\overline{H}}{\overline{H}_0} = c_2 + c_3 \frac{\overline{s}}{\overline{S}}, \qquad (4.26)$$

where the over-bars denote monthly average values, and $c_2 = 0.22$ and $c_3 = 0.54$ are determined empirically by Prescott (1940). Since then many empirical models have been developed that estimate global, direct, and diffuse radiation from the number of bright sunshine hours (Ahmad et al., 1991; Hay 1979; Iqbal 1979; Löf et al., 1966; Rietveld 1978; Şahin and Şen 1998). All these models utilize coefficients that are site specific and/or dependent on the averaging period considered. This confines their application to stations where the values of the coefficients were actually determined, or, at best, to localities of similar climate, and for the same average period.

Hay (1979) lessened the spatial and temporal dependence of coefficients by incorporating the effects of multiple reflections, but his technique requires surface and cloud *albedo* data. More recently, Suehrcke (2000) has argued that the relationship between global radiation and sunshine duration is approximately quadratic and thus the linear Eqs. 4.25 and 4.26 are of the wrong functional forms. A few authors have considered the relationships between sunshine duration, observed irradiation, and potential daily clear-sky beam radiation. Suehrcke and McCormick (1992) first proposed the following relationship:

$$\frac{\overline{H}_b}{\overline{H}_{bc}} = \frac{\overline{s}}{\overline{S}}, \qquad (4.27)$$

where $\overline{H}_b$ is the monthly averaged daily beam radiation on the horizontal surface, and $\overline{H}_{bc}$ is the monthly averaged potential daily clear-sky beam irradiation on a horizontal surface. The same relationship was subsequently used to predict the performance of a solar hot water system. Hinrichsen (1994) employed Eq. 4.25 to assign physical meaning to coefficients c_2 and c_3 in Eq. 4.26, while Suehrcke (2000) used Eq. 4.27 to derive his non-linear relationship between global radiation and sunshine duration.

The physical arguments suggest that the same relationship exists for irradiation at normal incidence. Indeed, Gueymard (1993) proposed that

$$\frac{\overline{H}_{bn}}{\overline{H}_{bnc}} = \frac{\overline{s}}{\overline{S}_c}, \qquad (4.28)$$

where $\overline{H}_{bn}$ is the monthly averaged daily beam irradiation at normal incidence, $\overline{H}_{bnc}$ is the monthly averaged potential daily clear-sky beam irradiation at normal incidence, and $\overline{S}_c$ is the monthly averaged day length modified to account for when the sun is above a critical solar elevation angle. The ratio $\overline{s}/\overline{S}_c$ is similar to $\overline{s}/\overline{S}$ in Eqs. 4.25–4.27 except $\overline{S}_c$ corrects for the irradiation threshold of sunshine recorders. The basis of Eqs. 4.27 and 4.28 is that for a given day the beam radiation incident at the surface (H_b or H_{bn}) is a fraction, s/S, of what would have been incident if the sky had been clear all day. In the absence of clouds, H_{bc} and H_{bnc} are functions of atmospheric scattering and absorption processes. The appeal of Eqs. 4.27 and 4.28 is twofold: they provide a means of estimating the potential beam irradiation and they do not contain empirically derived coefficients. However, a minimum averaging period is recommended when using these equations to estimate potential beam irradiation. A monthly period has been suggested by Gueymard (1993). The time averaging is necessary since s is simply the total number of sunshine hours per day and provides no information about when the sky was cloudless during any given day. There are several other assumptions in Eqs. 4.24 and 4.25. Turbidity and *precipitable water* are the same during cloudless and partly cloudy days, measurements of s are accurate, and the sunshine recorder threshold irradiance is constant and known.

4.4.1 Angström Model (AM)

Different global terrestrial solar radiation estimation models on the earth's surface are proposed, which use the sunshine duration data as the major predictor at a location. Some others include additional meteorological factors, such as the temperature and humidity, but all the model parameter estimations are based on the least squares technique and mostly a linear regression equation is employed for the relevant relationship between the terrestrial solar radiation and the predictor factors.

Angström (1924) provided the first *global solar radiation* amount estimation model from the *sunshine duration* data. This model expresses the ratio of the average global terrestrial irradiation, $\overline{H}$, to extraterrestrial irradiation, which is the cloudless irradiation, H_0, in linear relationship to the ratio of average sunshine duration, $\overline{S}$, to the cloudless sunshine duration, S_0, as

$$\frac{\overline{H}}{H_0} = a + b\frac{\overline{S}}{S_0}, \qquad (4.29)$$

with $a = 0.25$ and $b = 0.75$ for Stockholm, Sweden. According to historical records, in 1919 Kimball (1919) suggested the same idea and proposed $a = 0.22$ with $b = 0.78$. Later, Prescott (1940) modified this equation in such a manner that the summation $(a+b)$ is not equal to 1.0. He suggested that $a = 0.22$ and $b = 0.54$ and, hence, more realistic estimations are obtained. Physically, in Eq. 4.29 a corresponds to relative diffuse irradiation during overcast meteorological situations, whereas $(a+b)$ corresponds to the relative cloudless-sky condition global irradiation. An implied assumption in the structure of this linear model is the superposability of two extreme cloud states, which are reflected in the $(a+b)$ summation. However, in actual situations the superposability is not possible with respect to all possible combinations of atmospheric variables other than the cloud cover. This is the first indication why the summation $(a+b)$ did not equal to val1.0 as suggested by Prescott (1940). Furthermore, in practical applications, various non-linear estimation models are also proposed in order to relieve the assumption of superposability. Another physical fact that the solar radiation models should include non-linear effects is that atmospheric turbidity and turbulence in the planetary boundary do not necessarily vary linearly with total cloud cover. There are numerous studies and proposals as alternatives to the linear model in the solar energy literature and with the expectations of more studies in the future, but Gueymard et al. (1995) state that the studies related to solar radiation should now be more fully scrutinized. In particular, it is understood that the mere use of Angström's equation to estimate global irradiation from local sunshine data would generally be judged as not publishable unless a new vision in the model structure is documented. All these explanations indicate that linear models are very restrictive and, therefore, many researchers have tried to propose non-linear models for better refinements (Chap. 5).

The AM helps to estimate the amount of the global daily (H), monthly ($\overline{H}$), and yearly ($\overline{\overline{H}}$) solar radiation from the comparatively simple measurements of sunshine duration, S, according to

$$\frac{H}{H_0} = a + b\frac{S}{S_0}, \qquad (4.30)$$

where H_0 and S_0 are cloudless daily global irradiation received on a horizontal surface at ground level i.e., extraterrestrial and maximum possible sunshine duration; both a and b are model parameters. This equation has been used most often all over the world in order to calculate the global irradiation at locations of sunshine duration measurements and to extrapolate the global solar radiation estimations from measured short-term solar radiation data. Later, this equation has been modified by taking into account some other relevant meteorological variables (Abouzahr and Ramkumar 1991). All over the world, the coefficients are estimated from available solar radiation and sunshine duration data at a location by use of the statistical regression technique. However, in such an approach there are implied assumptions as follows (Şen 2001b):

1. The model parameters are assumed invariant with time on average, as if the same sunshine duration appears on the same days or months of the year in a particular location.
2. Whatever the scatter diagram of H versus S, automatically the regression line is fitted leading to constant a and b estimates for the given data. In fact, these coefficients depend on the variations in the sunshine duration during any particular time interval and since sunshine duration records have inherently random variability so are the model parameters.
3. Angström's approach provides estimations of the global solar radiation on horizontal surfaces, but, unfortunately, it does not give clues about global solar

radiation on a tilted surface because diffuse and direct irradiations do not appear in the AM.
4. The AM relates the global solar radiation to the sunshine duration only by ignoring the other meteorological factors such as the relative humidity, maximum temperature, air quality, latitude, and elevation above mean sea level. Each one of these factors contributes to the relationship between H and S and their ignorance causes some errors in the prediction and even in the model identification. For instance, Eq. 4.30 assumes that the global solar radiation on horizontal surfaces is proportional to the sunshine duration only. The effects of other meteorological variables always appear as deviations from the straight line fit on any scatter diagram all over the world. In order to overcome this, it is necessary to assume that the coefficients in Angström's equation are not constants but random variables that may change according to the capacity of measured data.
5. The physical meanings of the model coefficients are not considered in most of the application studies, but only the statistical linear regression line fit and parameter estimations are obtained directly and then incorporated into Eq. 4.30 for the global solar radiation estimation from the sunshine duration records. This is because the regression method does not provide dynamic estimation of the coefficients from available data.

There are also statistical restrictions in the parameter estimations of any regression technique based model such as the AM. These restrictions are as follows (Şen, 2001b):

1. *Linearity*: The regression technique fits a straight line trend through a scatter of data points and a correlation analysis tests for the "goodness-of-fit" of this line. Clearly, if the trend cannot be represented by a straight line, regression analysis will not portray it accurately. The unrestricted model described in Chap. 5 does not require such a restriction, since it is concerned with the variances and arithmetic averages only.
2. *Normality*: It is widely assumed that use of the regression model requires that the variables have normal distributions. The requirement is not that the raw data be normally distributed but that the conditional distribution of the residuals should be normally distributed. If the conditional distribution is normal, then it is almost certain that the distributions of global solar radiation and sunshine duration are also normally distributed. Thus, it is necessary to test if the data are normally distributed in order to inquire as to whether a necessary prerequisite for normal conditional distribution exists. The spatial arrays of monthly (H/H_0) data are generally not normally distributed. In the northern hemisphere from February to October the distributions are significantly positively skewed, suggesting that a few stations with especially large values of (H/H_0) produced a monthly mean that is greater than the mode. July, August, November, and December all displayed spatial distribution with significant negative kurtosis levels (Balling and Cerveny 1983).
3. *Means of conditional distributions*: For every value of sunshine duration, the mean differences between the measured and predicted global solar radiation

values obtained by Eq. 4.30 must be zero. If they are not, the coefficients of the regression equation (a and b) are biased estimates. The implication of major departure from this assumption is that the trend in the scatter diagram is not linear.

4. *Homoscedasticity*: It means equal variances in the conditional distributions and it is an important assumption. If it is not satisfied then the regression equation coefficients (a and b) may be severely biased (see Fig. 4.3b). In order to test for homoscedasticity the data must be subdivided into three or more groups and the variance of each group must be calculated. If there is significant difference between any of these variances then the data has homoscedasticity.

5. *Independence*: The crux of this assumption is that the value of each observation on the independent variable (sunshine duration in the AM model) is independent of all the other variables, so that one cannot predict the value of (S/S_0) at time, say i, if one knows (S/S_0) value at time, $i-1$. There are two interpretations as to the importance of this assumption, one is substantively logical and the other is statistically logical (Johnston 1980). The statistical interpretation of independence as a special case of autocorrelation relates to the linearity assumption.

6. *Lack of measurement error*: This assumption requires that both global solar radiation and sunshine duration measurements are without error. If this is not the case and the magnitude of the error is not known, then the coefficients of the regression equation may be biased to an extent that cannot be estimated.

If these six assumptions do not apply then the AM coefficient estimations may be under suspicion. The application of the unrestricted model in Sect. 4.6 does not require that the (H/H_0) versus (S/S_0) scatter diagram should have a distinguishable pattern as a straight line or a curve.

Many researchers (Ahmad *et al.*, 1991; Akinoğlu and Ecevit 1990; Angström 1924, 1929, 1956; Balling and Cerveny 1983; Barbaro *et al.*, 1978; Beckman *et al.*, 1977) have considered additional meteorological factors to Eq. 4.30 for the purpose of increasing the accuracy of the estimated values. This is equivalent to saying that deviations from the classic AM are explained with the additional variables. Although each one of these studies refined the coefficient estimates, they all depend on the average parameter values obtained by the least squares method and, therefore, there are still remaining errors although smaller than the original AM. Recently, a methodology has been presented that takes into consideration the random variations in the coefficients of the AM (Bucciarelli 1986).

The classic statistical analysis by the least squares technique and the regression method of Eq. 3.30 will lead to AM parameter estimations according to Eqs. 4.2 and 4.3 with the relevant notations as

$$b = \frac{\sum_{i=1}^{n}\left[\left(\frac{H}{H_0}\right)_i - \overline{\left(\frac{S}{S_0}\right)_i}\right]\left[\left(\frac{H}{H_0}\right)_{i-1} - \overline{\left(\frac{S}{S_0}\right)_{i-1}}\right]}{\sqrt{\sum_{i=1}^{n}\left[\left(\frac{H}{H_0}\right)_i - \overline{\left(\frac{S}{S_0}\right)_{i-1}}\right]^2 \left[\left(\frac{H}{H_0}\right)_{i-1} - \overline{\left(\frac{S}{S_0}\right)_{i-1}}\right]^2}} \quad (4.31)$$

and

$$a = \overline{\left(\frac{H}{H_0}\right)} - b\,\overline{\left(\frac{S}{S_0}\right)}. \tag{4.32}$$

However, these estimations are based on a set of restrictive regression assumptions, which will be explained below with the proposal of the unrestricted solar radiation model (Sect. 4.6).

4.5 Successive Substitution (SS) Model

Routinely recorded daily global irradiation and sunshine duration values are used by the regression technique for determining the coefficients as in Eqs. 4.31 and 4.32. Such a model provides unique estimations of global solar radiation given the sunshine duration. In order to consider the effects of the unexplained part, it is necessary to estimate coefficients from the successive data pairs "locally" rather than "globally" as in the AM approach (Şahin and Şen 1998).

Let us consider the physical and mathematical meanings of parameters a and b in Eq. 4.30. First of all, a represents the ratio of actual daily global irradiation, H, to the daily (or monthly) extraterrestrial irradiation, H_0, provided that physically the sun is covered by clouds all day, so that $S = 0$, i. e., overcast sky. On the other hand, b corresponds to the slope of the linear model, which is defined differentially as

$$b = \frac{d(H/H_0)}{d(S/S_0)}. \tag{4.33}$$

This first-order ordinary differential equation can be written in terms of the backward finite difference method as

$$b'_i = \frac{\left(\frac{H}{H_0}\right)_i - \left(\frac{H}{H_0}\right)_{i-1}}{\left(\frac{S}{S_0}\right)_i - \left(\frac{S}{S_0}\right)_{i-1}}, \quad (i = 2, 3, 4, \ldots, n). \tag{4.34}$$

Herein, n is the number of records and b'_i is the rate of local irradiation change with the sunshine duration between time instances, $i - 1$ and i. For daily data, these are successive daily rates of change or in the case of monthly records, monthly rates of change. Rearrangement of Eq. 4.30 and considering Eq. 4.34 leads to the successive time estimates of a'_i as

$$a'_i = \left(\frac{H}{H_0}\right)_i - b'_i \left(\frac{S}{S_0}\right)_i, \quad (i = 2, 3, 4, \ldots n). \tag{4.35}$$

The application of these last two equations to actual relevant data yields $n(n-1)/2$ coefficient estimations. Each pair of the coefficient estimate (a'_i, b'_i) explains the

4.5 Successive Substitution (SS) Model

whole information for successive pairs of global radiation and corresponding sunshine duration records. It has been already shown by Angström (1924) that physically a must be greater than zero. Comparisons of Eq. 4.34 with Eq. 4.31 and Eq. 4.35 with Eq. 4.32 indicate that the regression technique estimations do not allow any randomness in the calculation of coefficients.

It is possible to obtain the relative frequency distribution of the SS model coefficients in addition to statistical parameters such as the mean, variance, or standard deviation. Confidence limits on parameter estimations can also be stated at given significance levels such as 5% or 10%. Taking the average value of both sides in Eq. 4.35 leads to finite difference averages of the new Angström coefficients as

$$\overline{a'} = \overline{\left(\frac{H}{H_0}\right)} - \overline{b'}\overline{\left(\frac{S}{S_0}\right)}. \tag{4.36}$$

The difference of this expression from Eq. 4.32 results in

$$\overline{a'} - a = \left(b - \overline{b'}\right)\overline{\left(\frac{S}{S_0}\right)}. \tag{4.37}$$

The SS method is applied to a set of solar radiation and sunshine duration data measured at 29 sites (Table 4.3) in Turkey, which is located between latitudes 36°N and 42°N and longitudes 26°E and 45°E (Fig. 4.6). It has relatively significant solar energy potential especially in the southern parts including the Mediterranean Sea region.

At each station daily records are available concurrently for 25-year measurements, which are checked by plotting the extraterrestrial irradiation versus global irradiation and sunshine duration. Accordingly, extremely odd measurements are corrected through the classic regression technique.

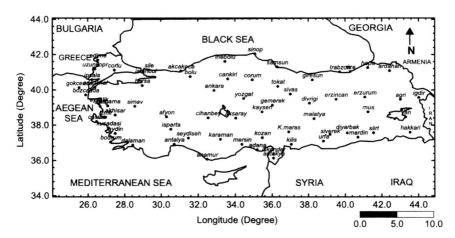

Fig. 4.6 Location map

The SS method is applied independently for each station and parameter estimation series are obtained for a'_i and b'_i. The lower order statistics for each station are shown in Table 4.4 together with the AM parameters, which give some idea of the standard deviation values.

It is to be noticed that in the application of the SS approach mode values are considered rather than arithmetic averages as in the AM. Also included in Table 4.4 is the relative error (RE) percentage between the AM arithmetic average and the mode values of the SS method. These percentages indicate how the arithmetic averages deviate from the mode values. By making use of the longitudes and latitudes from Table 4.3, the regional variations of the average parameters are shown in Fig. 4.7a,b which are obtained by the Kriging method, i.e., geostatistical methods in the computer (Journel and Huijbregts 1978). In this way, it is possible to obtain average a'_i and b'_i values for any location within the study area.

It is also obvious that both averages do not change significantly in the north-south direction, but east-west variations are more often. Coefficients have greater values in the eastern part of the country. This is meteorologically very plausible because

Table 4.3 Data characteristics

Station name	Latitude	Longitude	Elevation (m)
Adana	36.98	35.30	20
Adiyaman	37.75	38.28	678
Afyon	38.75	30.53	1034
Amasya	40.65	35.85	412
Anamur	36.10	32.83	5
Ankara	39.95	32.88	891
Antalya	36.88	30.70	51
Aydin	37.85	27.83	56
Balikesir	39.65	27.87	102
Bursa	40.18	29.07	100
Çanakkale	40.13	26.40	6
Çankiri	40.60	33.62	751
Diyarbakir	37.92	40.20	677
Elazig	38.67	39.22	991
Erzincan	39.73	39.50	1218
Eskisehir	39.77	30.52	789
İstanbul	40.97	29.08	399
Isparta	37.77	30.55	997
İzmir	38.40	27.17	25
Kars	40.60	43.08	1775
Kastamonu	41.37	33.77	800
Kayseri	38.72	35.48	1093
Kirsehir	38.13	34.17	985
Konya	38.87	32.50	1031
Malatya	38.35	38.30	898
Mersin	36.82	34.60	5
Samsun	41.28	36.33	44
Trabzon	41.00	39.72	30
Van	38.47	43.35	16.71

4.5 Successive Substitution (SS) Model

Table 4.4 Statistical characteristics

Station name	AM		SS Mode		SD		Relative error (%)	
	a	b	a'	b'	a'	b'	a, a'	b, b'
Adana	0.33	0.29	0.31	0.31	1.27	1.75	6.05	8.42
Adiyaman	0.30	0.22	0.27	0.26	0.80	1.10	9.93	17.90
Afyon	0.40	0.28	0.34	0.29	0.78	1.95	15.20	5.78
Amasya	0.30	0.38	0.27	0.37	2.44	6.79	8.72	3.92
Anamur	0.36	0.25	0.29	0.35	1.51	2.04	21.40	28.90
Ankara	0.31	0.32	0.30	0.48	3.03	6.09	4.74	32.00
Antalya	0.33	0.38	0.33	0.32	1.98	2.78	0.60	16.20
Aydin	0.32	0.42	0.33	0.42	1.88	3.38	4.50	0.95
Balikesir	0.23	0.37	0.22	0.34	1.39	3.98	0.88	6.28
Bursa	0.27	0.33	0.24	0.35	1.25	3.09	9.62	4.03
Çanakkale	0.31	0.33	0.31	0.45	3.81	4.58	2.50	27.70
Çankiri	0.35	0.32	0.11	0.25	5.92	8.86	57.30	21.60
Diyarbakir	0.23	0.48	0.43	0.42	2.09	3.06	45.00	12.73
Elazig	0.32	0.32	0.24	0.40	1.74	2.46	25.40	18.38
Erzincan	0.44	0.15	0.40	0.25	0.83	2.42	9.15	38.05
Eskişehir	0.39	0.26	0.34	0.43	0.74	1.56	13.81	39.60
İstanbul	0.30	0.35	0.28	0.55	0.90	3.34	5.08	35.50
Isparta	0.36	0.16	0.28	0.16	0.58	1.03	21.32	0.00
İzmir	0.33	0.33	0.32	0.42	1.16	1.71	1.80	22.00
Kars	0.50	0.12	0.74	0.41	1.32	2.22	33.30	70.00
Kastamonu	0.32	0.24	0.19	0.31	0.70	2.04	41.17	21.20
Kayseri	0.36	0.23	0.31	0.30	2.87	4.02	13.73	24.10
Kirsehir	0.43	0.20	0.18	0.25	1.31	2.15	57.30	21.81
Konya	0.38	0.27	0.31	0.39	1.79	3.80	20.00	32.20
Malatya	0.31	0.37	0.24	0.47	2.04	0.47	23.45	21.80
Mersin	0.33	0.40	0.27	0.48	0.87	1.25	18.60	17.30
Samsun	0.34	0.31	0.22	0.40	2.78	8.23	33.70	22.40
Trabzon	0.28	0.38	0.26	0.46	8.82	23.69	8.27	17.86
Van	0.51	0.14	0.40	0.23	34.07	41.92	21.76	38.60

eastern Turkey has many days of the year where the sky is overcast due to high elevations reaching up to 6500 m above the sea level. The cold period of the year lasts almost 9 months. In contrast, from the Aegean Sea coast in the west toward central Anatolia, days with a cloudless sky are encountered much more frequently. In Fig. 4.7b the regional variation of $\overline{b'}$ coefficient has just the opposite trend to the east-west variation of $\overline{a'}$ and this is the expected meteorological situation as explained above. In order to support these conclusions in detail, Fig. 4.8 and 4.9 show the regional variations of the same parameters but this time with elevation instead of longitude as the abscissa.

Figures 4.7–4.9 relate the geographical (regional, spatial) information to solar radiation variations over the country. Figure 4.10 shows the relation between $\overline{a'}$ and $\overline{b'}$ for values calculated for all over Turkey.

There appears an inverse relationship between these two parameters. It means that low $\overline{b'}$ values are coupled with high $\overline{a'}$ values and *vice versa*. The general trend

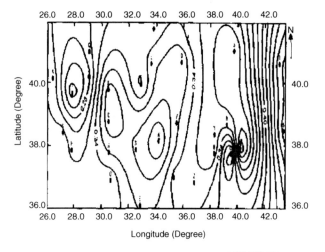

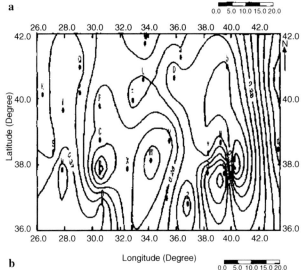

Fig. 4.7 a,b. Parameter regionalization

relationship between $\overline{a'}$ and $\overline{b'}$ is

$$\overline{a'} = -0.61\overline{b'} + 0.52 \,. \tag{4.38}$$

The empirical relative frequency distributions of average $\overline{a'}$ and $\overline{b'}$ values on a regional basis are shown in Fig. 4.11. It is to be noticed from Table 4.4 that $\overline{a'}$ and $\overline{b'}$ values are confined between zero and one. This piece of information implies that a beta-type of theoretical probability distribution function (PDF) is suitable for the parameters regional estimation (Bucciarelli 1984). Figure 4.11 shows both observed and beta-type PDF for $\overline{a'}$ and $\overline{b'}$.

4.5 Successive Substitution (SS) Model 125

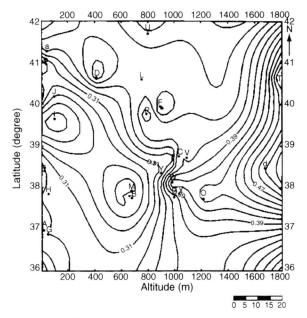

Fig. 4.8 Parameter map for $\overline{a'}$

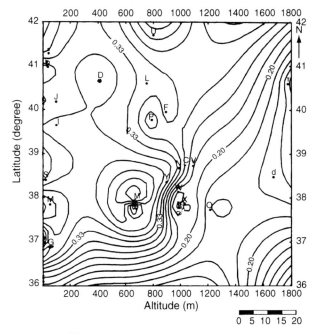

Fig. 4.9 Parameter map for $\overline{b'}$

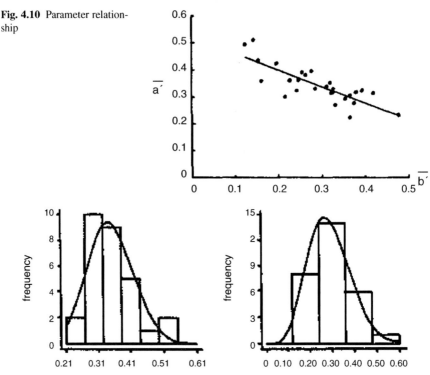

Fig. 4.10 Parameter relationship

Fig. 4.11 Distribution of $\overline{a'}$ and $\overline{b'}$ over Turkey

The beta distribution is checked to fit the empirical data with the chi-square criterion in the statistics literature at the 5% significance level. It is obvious even visually from Fig. 4.11 that the beta-type PDF is very convenient. The beta-type curves are helpful for any modeling studies in the future for areal radiation variations (see App. A).

4.6 Unrestricted Model (UM)

In practice, the estimation of model parameters is achieved most often by the least squares method and regression technique using procedural restrictive assumptions as already mentioned in Sect. 4.4.1. These are unnecessary procedural restrictions, which lead to unreliable biases in the parameter estimations. The averages and variances of the solar radiation and sunshine duration data play a predominant role in many calculations and the conservation of these parameters becomes more important than the cross-correlation coefficient. Gordon and Reddy (1988) stated that a simple functional form for the stationary relative frequency distribution for daily solar radiation requires knowledge of the mean and variance only. In almost any es-

4.6 Unrestricted Model (UM)

timate of solar radiation by means of computer software, the parameter estimations are achieved without caring about the theoretical restrictions in Sect. 4.4.1. The application of the regression technique to Eq. 4.30 for estimating the model parameters from the available data (Bucciarelli 1986) leads to

$$b = r_{hs} \sqrt{\frac{Var(H/H_0)}{Var(S/S_0)}} \qquad (4.39)$$

and

$$a = \overline{\left(\frac{H}{H_0}\right)} - r_{hs} \sqrt{\frac{Var(H/H_0)}{Var(S/S_0)}} \overline{\left(\frac{S}{S_0}\right)}, \qquad (4.40)$$

where r_{hs} is the cross-correlation coefficient between global solar radiation and sunshine duration data, $Var(\ldots)$ is the variance of the argument, and the over-bars indicate arithmetic averages. As a result of the classic regression technique, the variance of the predictand, given the value of predictor is

$$Var\left[\overline{(H/H_0)}/\overline{(S/S_0)} = S/S_0\right] = \left(1 - r_{rs}^2\right) Var\overline{(H/H_0)}. \qquad (4.41)$$

This expression provides the mathematical basis for interpreting r_{rs}^2 as the proportion of variability in $\overline{(H/H_0)}$ that can be explained provided that $\overline{(S/S_0)}$ is given. From Eq. 4.41, after rearrangement, one can obtain

$$r_{rs}^2 = \frac{Var\overline{(H/H_0)} - Var\left[\overline{(H/H_0)}/\overline{(S/S_0)} = S/S_0\right]}{Var\overline{(H/H_0)}}. \qquad (4.42)$$

If the second term in the numerator is equal to zero, then the regression coefficient will be equal to one. This is tantamount to saying that given $\overline{S/S_0}$ there is no variability in $\overline{(H/H_0)}$. Similarly, if it is assumed that $Var\left[\overline{(H/H_0)}/\overline{(S/S_0)} = (S/S_0)\right] = Var\overline{(H/H_0)}$, then the regression coefficient will be zero. This means that given $\overline{(S/S_0)}$ the variability in $\overline{(H/H_0)}$ does not change. In this manner, r_{rs}^2 can be interpreted as the proportion of variability in $\overline{(H/H_0)}$ that is explained by knowing $\overline{(S/S_0)}$. The requirement of normality is not satisfied, especially if the period for taking averages is less than one year. Since, daily or monthly data are used in most practical applications it is over-simplification to expect marginal or joint distributions to abide by the Gaussian (normal) PDF.

The UM parameter estimations require two simultaneous equations since there are two parameters to be determined. The average and the variance of both sides in Eq. 4.30 lead without any procedural restrictive assumptions to the following equations (Şen 2001b):

$$\overline{\left(\frac{H}{H_0}\right)} = a_u + b_u \overline{\left(\frac{S}{S_0}\right)} \qquad (4.43)$$

and

$$Var\left(\overline{H/H_0}\right) = b_u^2 Var\left(\overline{S/S_0}\right), \qquad (4.44)$$

where for distinction the UM parameters are shown as a_u and b_u, respectively. These two equations are the basis for the conservation of the arithmetic mean and variances of global solar radiation and sunshine duration data. The basic AM remains unchanged whether the restrictive or unrestrictive model is used. Equation 4.43 implies that in both models the centroid, i. e., averages of the solar radiation and sunshine duration data are equally preserved. Furthermore, another implication from this statement is that the AM and the UM yield close estimations around the centroid. The deviations between the two model estimations appear at solar radiation and sunshine duration data values away from the arithmetic averages. The simultaneous solution of Eqs. 4.43 and 4.44 yields parameter estimates as

$$b_u = \sqrt{\frac{Var\left(\overline{H/H_0}\right)}{Var\left(\overline{S/S_0}\right)}} \qquad (4.45)$$

and

$$a_u = \overline{\frac{H}{H_0}} - \sqrt{\frac{Var\left(\overline{H/H_0}\right)}{Var\left(\overline{S/S_0}\right)}} \overline{\left(\frac{S}{S_0}\right)}. \qquad (4.46)$$

Physically, variations in the solar radiation data are always smaller than the sunshine duration data and, consequently, $Var\left(\overline{S/S_0}\right) \geq Var\left(\overline{H/H_0}\right)$. For Eq. 4.45, this means that always $0 \leq b_u \leq 1$. Furthermore, Eq. 4.45 is a special case of Eq. 4.39 when $r_{hs} = 1$. The second term in Eq. 4.46 is always smaller than the first one, and hence $a_u > 0$. The following relationships are valid between the AM and UM parameters:

$$b_u = \frac{b}{r_{hs}} \qquad (4.47)$$

and

$$a_u = \frac{a}{r_{hs}} + \left(1 - \frac{1}{r_{hs}}\right)\overline{\left(\frac{H}{H_0}\right)}. \qquad (4.48)$$

According to Eq. 4.47, $b_u > b$ since always $0 \leq r_{hs} \leq 1$. Figure 4.12 indicates the straight line from the AM and the UM for the same data.

4.6 Unrestricted Model (UM)

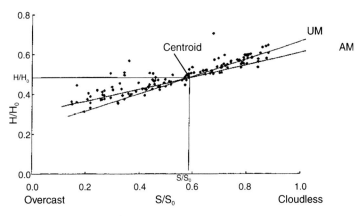

Fig. 4.12 Solar energy data centroidy

This further indicates that in the light of the previous statement the UM overestimates (under-estimates) compared to the AM estimations for sunshine duration data greater (smaller) than the average value. On the other hand, Eq. 4.48 shows that $a_u < a$. Furthermore, the summation of model parameters gives

$$a_u + b_u = \frac{a+b}{r_{hs}} + \left(1 - \frac{1}{r_{hs}}\right)\overline{\left(\frac{H}{H_0}\right)}. \tag{4.49}$$

These last expressions indicate that the two approaches are completely equivalent to each other for $r_{hs} = 1$. The application supposes that the AM is first used to obtain a_u, b_u, and r_{hs}. If r_{hs} is close to 1, then the AM coefficients estimation with restrictions is almost equivalent to a_u and b_u. Otherwise, the UM results should be considered for applications.

For the implementation of UM and AM parameter estimations, 29 global solar radiation stations are considered as given in Table 4.5.

Table 4.5 AM and UM parameters

Station name	AM		UM	
	a	b	a_u	b_u
Adana	0.33	0.29	0.20	0.50
Adiyaman	0.30	0.22	0.25	0.28
Afyon	0.40	0.28	0.36	0.34
Amasya	0.30	0.38	0.26	0.41
Anamur	0.36	0.25	0.26	0.40
Ankara	0.31	0.32	0.28	0.38
Antalya	0.33	0.38	0.22	0.55
Aydin	0.32	0.42	0.25	0.53
Balikesir	0.23	0.37	0.20	0.41
Bursa	0.27	0.33	0.21	0.46
Çanakkale	0.31	0.33	0.27	0.41
Çankiri	0.35	0.32	0.32	0.40

Table 4.5 (continued)

Station name	AM		UM	
	a	b	a_u	b_u
Diyarbakir	0.23	0.48	0.16	0.61
Elazig	0.26	0.46	0.18	0.52
Erzincan	0.44	0.15	0.34	0.33
Eskişehir	0.39	0.26	0.33	0.39
İstanbul	0.30	0.35	0.27	0.41
Isparta	0.36	0.16	0.30	0.26
İzmir	0.33	0.33	0.25	0.45
Kars	0.50	0.12	0.34	0.45
Kastamonu	0.32	0.24	0.29	0.29
Kayseri	0.36	0.23	0.30	0.36
Kirsehir	0.43	0.20	0.36	0.30
Konya	0.38	0.27	0.30	0.40
Malatya	0.31	0.37	0.26	0.45
Mersin	0.33	0.40	0.36	0.45
Samsun	0.31	0.23	0.30	0.31
Trabzon	0.28	0.38	0.23	0.51
Van	0.51	0.14	0.38	0.34

In order to see whether the global solar radiation and sunshine duration data are normally distributed, the frequency distribution function (FDF) of the ratios are plotted for the months of January, April, July, and October in Figs. 4.13 and 4.14.

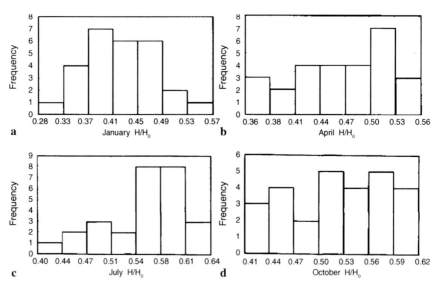

Fig. 4.13 a–d. Areal frequency distribution functions of H/H_0 in **a** January, **b** April, **c** July, and **d** October

4.6 Unrestricted Model (UM)

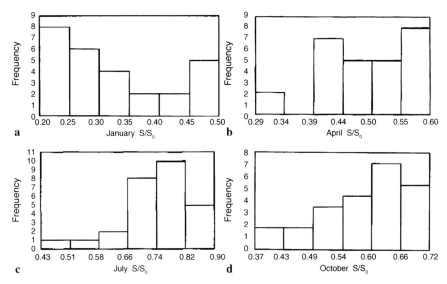

Fig. 4.14 a–d. Frequency distribution functions of S/S_0 **a** January, **b** April, **c** July, and **d** October

It is obvious that FDFs are not Gaussian (normal) and they are overwhelmingly skewed. Hence, the AM parameter estimations by the classic least squares technique remain biased. By making use of the AM and UM parameter estimates, their FDFs over Turkey are presented in Fig. 4.15.

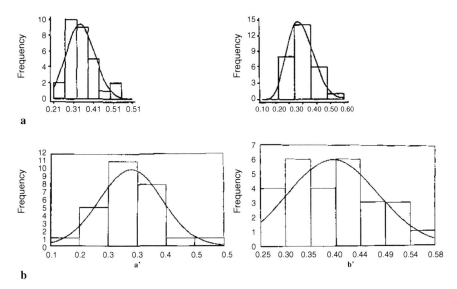

Fig. 4.15 a,b. Comparison of areal parameter FDFs of **a** AM and **b** UM parameter estimations

This figure indicates that both methods yield FDFs close to a normal function. It is obvious that the UM function provides over-estimates compared with the regression results for high values of sunshine duration data that are greater than the average sunshine duration. This point may be a reasonable answer to the statement by Gueymard *et al.* (1995) that for clear-sky conditions the AM under-estimates insulation. In Fig. 4.16 the AM and UM approaches are presented for the Ankara, Adana, and İstanbul stations. It is noticed that both straight lines pass through the centroid $(\overline{H/H_0}, \overline{S/S_0})$ of the scatter diagram.

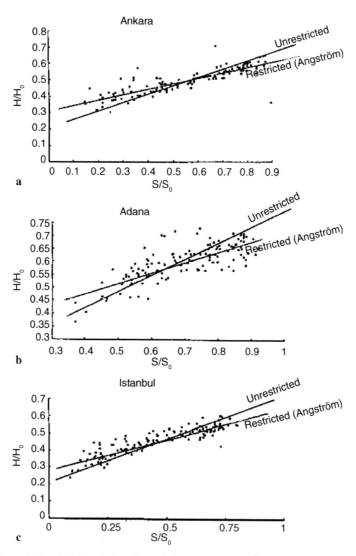

Fig. 4.16 a–c. AM and UM for Ankara (*top*), Adana (*middle*), and İstanbul (*bottom*)

4.7 Principal Component Analysis (PCA) Model

One of the practical observations by visual inspection of a scatter diagram is the impression of reliability in the application of classic statistical methods such as the regression line or curve. The more the scatter the less reliable are the conclusions because the regression models try to represent the data and the dependence between the data constituents on average without giving preference to the variability within each set of data. Although each factor has its own variance, they can be comparable with each other, but their joint behavior on a scatter diagram represents definitely different variances along different directions. Among these directions, there is one with the maximum joint variance and another one perpendicular to this direction with the minimum joint variance. Here, joint variance means the variance of scatter point projections along any desired direction as shown in Fig. 4.17.

Herein, the I-I direction is an axis within the scatter diagram and the projections of each scatter point on this axis provides a new data set, which is composed of partial sunshine duration and partial solar radiation data. In order to fix the position of such an axis, the centroid of the original data (arithmetical averages, $\overline{S/S_0}$ and $\overline{H/H_0}$) are considered with a rotation angle α of the new axis from the horizontal axis. For any given set of solar radiation and sunshine duration data, change in the rotation angle causes changes in the projection data, and accordingly their variances also. The changes of projection data variances with α and the axis perpendicular to it $(\alpha + \pi/2)$ appear in the forms of harmonic oscillations, where there is one rotational angle with maximum variance of projection data and another one perpendicular to

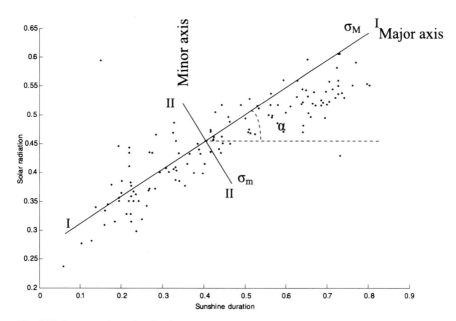

Fig. 4.17 Scatter and rotational axis

this axis with the minimum variance (II-II). In such a situation, the following new definitions become valid:

1. The rotational angle attached to the maximum variance axis is referred to as the *principal rotation* angle, α_p.
2. The axis with the maximum projection data variance, σ_M^2, is the major *principal axis*, which explains the maximum joint variance direction within the solar radiation and sunshine measurement.
3. The axis with the minimum projection data variance, σ_m^2, is the minor principal axis with the next minimum joint variance direction within the solar radiation and sunshine measurement.

The projected data on the principal axes are independent from each other. This means to say that the original dependent data are transformed into two independent data sets along the principal axes. Such a property provides convenience in modeling the projected data sets within themselves independently from other variables and therefore, their averages and variances become important since the cross-correlation coefficient between these projected data sets (along I-I and II-II) is equal to zero. These principal axes are the basis of principal component analysis (PCA) (Davis 1986).

The two dimensional PCA preserves the total variance of the two variables, namely, S/S_0 and H/H_0. The summation of solar radiation variance σ_{SI}^2 and the sunshine duration variance σ_{SD}^2 is equal to the summation of the principal variances. This can be expressed as follows:

$$\sigma_{SI}^2 + \sigma_{SD}^2 = \sigma_M^2 + \sigma_m^2 . \qquad (4.50)$$

This expression is obvious from Fig. 4.17 since the summation of variances at any rotation angle remains constant and equal to the summation of the basic variances. On the other hand, by definition always $\sigma_M^2 > \sigma_m^2$, and accordingly the ratio is

$$r = \sigma_m^2/\sigma_M^2 < 1 , \qquad (4.51)$$

which can be used as a measure of dependence between the basic factors of solar radiation and sunshine duration. In order to indicate the scatter diagram variations along this principal axis, it is possible by the use of principal variances to show the *elliptical variation*s. The slimmer the ellipse the more is the dependence between the solar radiation and sunshine duration data (see Fig. 4.18).

The smaller the ratio the more dependent are the two variables, and practically, if $r < 0.05$ then the minor principal axis can be neglected in the calculations, and the whole variability is considered to be accounted for by the major principal variable only. It is possible to deduce that as $r \to 0$ the major principal axis variable becomes more dominant, and hence the application of the AM or the UM is tractable mathematically and plausible physically. This is exactly what the linear regression is, even though it may not be suitable physically. As $r \to 1$, the two principal variances

4.7 Principal Component Analysis (PCA) Model

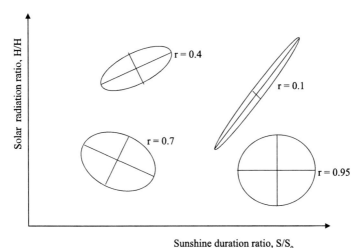

Fig. 4.18 Projection variance and rotation angle

become equal to each other and this implies the most scatter of the solar radiation and sunshine duration points. In this case, the use of the classic regression equation cannot be valid. Since $0 < r < 1$, the use of two-dimensional PCA becomes more suitable than any regression technique (Davis 1986). In PCA theory, the principal axes are referred to as the eigenvectors of the correlation matrix and the variances are the eigenvalues (Cressie 1993; Davies and McKay 1989). In two-dimensional PCA the original data values, say $(S/S_0)_i$ and $(H/H_0)_i$, are transformed to principal variables p_{1i} and p_{2i} through the axis rotation formulation as

$$p_{1i} = \left(\frac{H}{H_0}\right)_i \cos\alpha - \left(\frac{S}{S_0}\right)_i \sin\alpha \qquad (4.52)$$

and

$$p_{2i} = \left(\frac{H}{H_0}\right)_i \sin\alpha + \left(\frac{S}{S_0}\right)_i \cos\alpha . \qquad (4.53)$$

These principal variables are the projections of the solar radiation-sunshine duration data pairs on the principal axes. If the second component is comparatively smaller than the first one, i.e., $r < 0.05$ then the only variable that represents the two original variables is the first principal value. In the same way, the inverse transformation can be written trigonometrically as

$$\left(\frac{S}{S_0}\right)_i = p_{1i} \sin\alpha + p_{2i} \cos\alpha \qquad (4.54)$$

and

$$\left(\frac{H}{H_0}\right)_i = -p_{1i} \cos\alpha - p_{2i} \sin\alpha. \qquad (4.55)$$

Five representative solar radiation and sunshine duration measurement sites are considered for application of PCA from different climatology regions of Turkey, which lie within the northern subtropical climate region of the world with four distinctive seasons. This location is in the transitional region between subtropical and polar regional weather and climatic features. The five measurement locations, İstanbul, Ankara, Antalya, Trabzon, and Kars, are indicated in Fig. 4.6 with statistical parameters of each site as in Table 4.6.

İstanbul lies between the borders of two continents, Europe and Asia. The Bosphorus connects the Black Sea to the Mediterranean Sea through the inland sea of Marmara and then the Aegean Sea (Fig. 4.6). The city has a modified Mediterranean climate in the summer, but in winter the cold weather movements from central Siberia in Russia and cold wave movements from the Balkan Peninsula in eastern Europe cause rather cold spells.

The Ankara station lies within the central Anatolian plateau with steppe features and dry climatic conditions. The climatic signature is continental with dry and hot summers. However, in winter the weather is very cold.

Antalya is under the influence of the Mediterranean climate with hot summers and mild winters. Since this site lies within the coastal plane, the solar radiation values are very high in summer and rather low in the winter periods. The solar radiation is under the influence of humidity due to excessive evaporation in this region.

Trabzon lies along the eastern Black Sea coast and receives much more rainfall than the other sites mentioned in this section. Due to its location, the summers are humid but winters are dry and cold.

Kars is located in the eastern mountainous part of Turkey with altitudes reaching to almost 2,500 m above mean sea level. The climate is strongly continental with very short summers of about two months at the maximum and long winter periods. The weather fluctuation is very frequent in the region leading to randomly varying sunshine duration hours within each day, especially, during the summers.

The application of two-dimensional PCA to solar radiation and sunshine duration data at these stations shows different elliptical variations. Table 4.7 presents the rotation angle and the principal axis variances in addition to the *variance ratio, r*.

It is obvious from this table that the maximum rotation angles appear in Antalya in the south and in Trabzon in the north. The increase in the rotation axis angle implies that the site is subject to rather stable weather conditions. The greater is the

Table 4.6 Site location features

Station name	Solar radiation		Sunshine duration	
	Average	Variance	Average	Variance
İstanbul	0.4549	0.0065	0.4494	0.0392
Ankara	0.4894	0.0061	0.5496	0.0430
Antalya	0.6014	0.0057	0.7045	0.0188
Trabzon	0.4043	0.0030	0.3330	0.0113
Kars	0.5651	0.0063	0.5051	0.0307

4.7 Principal Component Analysis (PCA) Model

Table 4.7 Two-dimensional PCA parameters

Station name	Rotation angle (degree)	Principal axis variances Major (10^{-2})	Minor (10^{-3})	Variance ratio
İstanbul	20	4.42	1.48	0.033
Ankara	19	4.76	1.50	0.031
Antalya	24	2.20	2.53	0.115
Trabzon	23	1.31	1.21	0.092
Kars	8	3.12	5.79	0.185

rotation angle, the more stable is the weather situation, and hence, the solar radiation variation is not very much with time.

As a first approximation the major principal axis is considered as the model that helps to predict the solar radiation amounts from a given set of sunshine duration data. In such a modeling the effect of the minor axis and, therefore, the variation along this axis are ignored. Similar to the AM, the straight line equation of the major principal axis can be considered as the valid model. The intercept and slope parameters corresponding to the a and b parameters in Eq. 4.30 are calculated and shown as a_p and b_p in Table 4.8 together with the AM and UM parameters.

In Fig. 4.19 the three straight lines resulting from the three models are shown and they all pass through the data centroid and, consequently, they all predict the solar radiation data from the sunshine duration with almost equal efficiencies around the arithmetic means of the sunshine duration values.

However, as the location of sunshine duration value moves away from the arithmetic average region, the estimation by the UM becomes an over-estimate of the solar radiation values. In such situations the best model appears to be the two-dimensional PCA, which is shown by the dotted line in Fig. 4.19. It is obvious that the PCA method gives rather closer results in each case to the classic AM, which has restrictive assumptions as already explained in the previous section; the PCA method does not have such restrictions. It is, therefore, recommended to employ PCA in solar radiation and sunshine duration data modeling and in solar radiation prediction studies.

Table 4.8 Parameters of different models

Station name	AM		UM		PCA	
	a	b	a_u	b_u	a_p	b_p
İstanbul	0.297	0.351	0.273	0.393	0.2913	0.3640
Ankara	0.312	0.323	0.376	0.355	0.300	0.3443
Antalya	0.332	0.382	0.372	0.496	0.288	0.4452
Trabzon	0.279	0.376	0.139	0.489	0.2629	0.4245
Kars	0.508	0.113	0.334	0.384	0.4941	0.1405

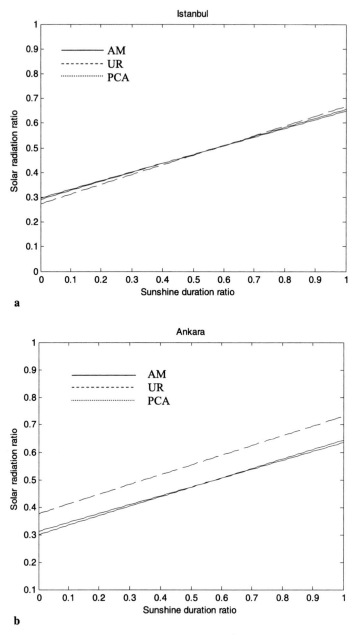

Fig. 4.19 a,b. Solar radiation-sunshine duration models for **a** İstanbul, **b** Ankara

4.7 Principal Component Analysis (PCA) Model

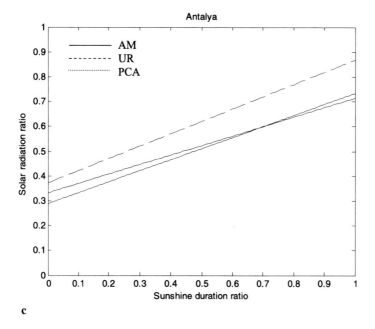

c

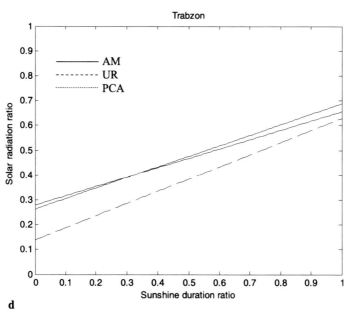

d

Fig. 4.19 c,d. Solar radiation-sunshine duration models for **c** Antalya, **d** Trabzon

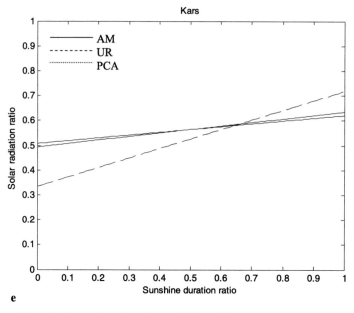

Fig. 4.19 e. Solar radiation-sunshine duration models for e Kars

4.8 Linear Cluster Method (LCM)

Solar radiance estimation from measurable sunshine duration data constitutes one of the fundamental engineering problems in solar energy applications. The general relationship including various astronomical and meteorological factors can be written mathematically as

$$\frac{H}{H_0} = f\left(\frac{S}{S_0}, h, T, r, \delta, \phi\right), \qquad (4.56)$$

where h is the elevation, T and r are the meteorological parameters of temperature and relative humidity, and δ and ϕ are the astronomical parameters of declination angle and latitude, respectively. The explicit form of Eq. 4.56 is presented by Gopinathan (1986) as a multiple linear regression expression

$$\frac{H}{H_0} = a + b\frac{S}{S_0} + ch + dT + er + f\delta + g\phi, \qquad (4.57)$$

where a, b, c, d, e, f, and g are the model parameters. The first two terms on the right-hand side express the AM. In practice, most often without looking at the scatter diagrams between the variables a linear model is assumed and the deviations from such a model are not considered in detail for the model suitability. Many linear relationships have been critically reviewed by Gueymard *et al.* (1995) who

4.8 Linear Cluster Method (LCM)

suggested that AMs are overwhelmingly dominant in the literature and, therefore, future researchers should be directed toward more physically based methods.

Şen and Öztopal (2003) have presented a *linear cluster model* (LCM) for the estimation of solar radiation. In the LCM rather than plotting the scatter diagram with the same symbol for all the data without distinction between seasons, data for each season are plotted by different symbols. This helps to identify and interpret different subgroups or, as in many cases, the interference between different seasons can be appreciated with meaningful astronomical and meteorological interpretations. In such an identification problem, the following points are the most significant to keep in mind for interpretations:

1. Physically, different seasons are expected to cluster in different groups. If only astronomical effects are assumed to play a role in the arrival of solar radiation onto the earth's surface, without any atmospheric and meteorological effects, then they appear in distinctive non-overlapping groups. However, meteorological factors cause interferences between the seasons.
2. In the case of astronomic effects only, the H/H_0 versus S/S_0 plot is expected to appear in the form of a straight line with the assumption of a linear model as shown in Fig. 4.20. Here, $(H/H_0)_m$ and $(H/H_0)_M$ represent the minimum and maximum solar radiation ratios, respectively. Similarly, on the horizontal axis $(S/S_0)_m$ and $(S/S_0)_M$ are the astronomically possible minimum and maximum values of sunshine duration, respectively. The linear model appears as a straight line on such a graph without seasonal distinctions. The minimum (maximum) irradiation and sunshine duration are expected logically to appear at the winter (summer) *solstice*. The *equinox* dates appear between these two extreme cases.
3. Consideration of seasonal distinction provides four non-overlapping parts along the straight line as shown in Fig. 4.21. It is possible to interpret logically that the lowest part of the straight line corresponds to the winter period because during this season the solar radiation has its lowest values in the northern hemisphere. The highest part represents summer, and in between the lower part corresponds to autumn and the upper part to spring, respectively.
4. Once the meteorological conditions and atmospheric effects start to play a role, over many years, within each month, the solar irradiance assumes values depending on the sunshine duration variations, which can be clear, overcast, or cloudy. As a result of the atmospheric composition, as particulate matter, CO_2, water vapor *etc.* change in a random manner, the astronomically linear parts become blurred, vague, randomly scattered, or fuzzy, which are different sources of uncertainties. For instance, such an uncertain scatter of points for the winter period only is shown in Fig. 4.22. It is necessary to notice that the scatter of points around the astronomical part appears not only above or below the straight line, but also interferes with the adjacent season.
5. In extreme weather, in some meteorological and atmospheric situations, there are interference scatters not only between adjacent seasons but occasionally also between non-adjacent seasons also. Figure 4.23 shows such a general conceptual scatter diagram. However, the interferences between adjacent seasons are more intense than those between non-adjacent ones. Scatter diagrams similar to

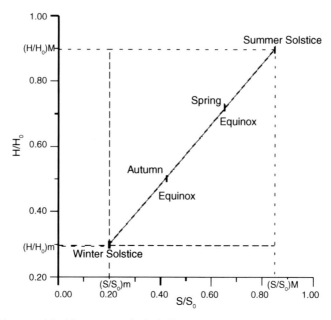

Fig. 4.20 Linear model without meteorological effects

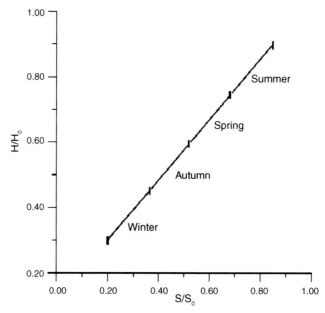

Fig. 4.21 Linear model with seasonal distinction

Fig. 4.23 lead to trends of generally linear and occasionally non-linear models without distinction between seasonal effects.

4.8 Linear Cluster Method (LCM)

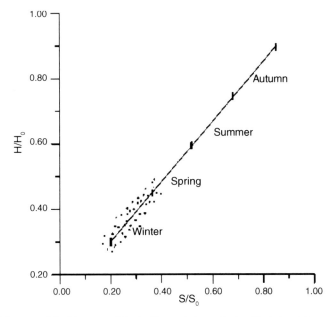

Fig. 4.22 Linear model with seasonal distinction and atmospheric effects in winter

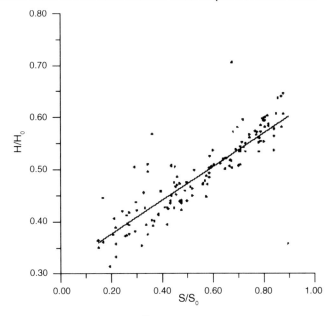

Fig. 4.23 Conceptual linear model scatter diagram

The application of the LCM is presented for two different solar radiation measurements from different climatic regions of Turkey, namely, İstanbul and Ankara stations. Since the solar radiation has seasonal and astronomical backgrounds, it is useful to sub-group the global scatter into seasonal or bi-seasonal groups. In Figs. 4.24 and 4.25 the best possible groupings for İstanbul and Ankara stations are shown, respectively.

It is possible to observe seasonal interferences from these graphs in addition to the seasonal and bi-seasonal scatters. The seasonal and bi-seasonal applications of the AM led to the parameter estimations through the classic regression technique, as shown in Table 4.9.

Comparison of this table with the global AM parameters in Table 4.5 indicates significant seasonal deviations between the corresponding parameters. In any efficient calculation of the solar energy amounts in Turkey, the winter season data and corresponding parameter estimates must not be considered at all because, in practice, solar heaters are not functional in Turkey during this season. In fact, similar arguments are also valid to a certain extent for the autumn season. Inclusion of the winter season in the global parameter estimations as presented in Table 4.5 imbeds a certain bias, which affects realistic solar energy calculations. Most often, in many parts of Turkey, the spring–summer bi-seasons are promising for sunshine duration and, hence, solar energy. Consequently, in any solar energy design and calculations, the parameters required are in the last two columns of Table 4.9.

Table 4.9 LCM coefficients

Station	Season								Bi-seasonal			
	Autumn		Winter		Spring		Summer		Autumn–winter		Spring–summer	
	a	b	a	b	a	b	a	b	a	b	a	b
İstanbul	0.54	0.02	0.28	0.45	0.23	0.49	0.52	0.05	0.30	0.36	0.29	0.36
Ankara	0.37	0.24	0.30	0.35	0.25	0.40	0.18	0.49	0.33	0.31	0.22	0.44

4.8 Linear Cluster Method (LCM)

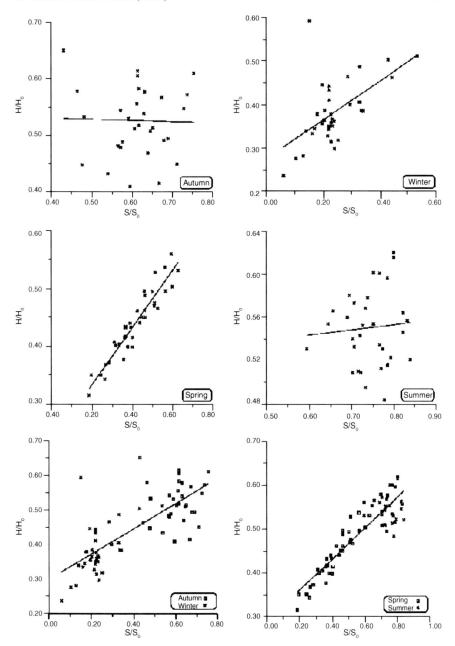

Fig. 4.24 Seasonal and bi-seasonal data scatter for İstanbul

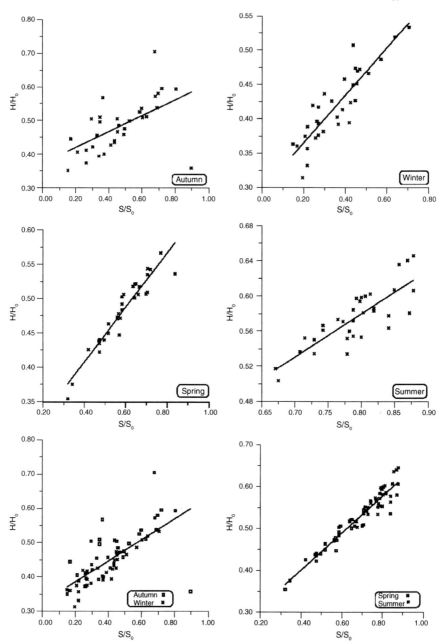

Fig. 4.25 Seasonal and bi-seasonal data scatter for Ankara

References

Abouzahr I, Ramkumar R (1991) Loss of power supply probability of stand-alone photovoltaic systems. IEEE Trans Energy Convers 6:1–11

Ahmad FA, Burney SM, Husain SA (1991) Monthly average daily global beam and diffuse solar radiation and its correlation with hours of bright sunshine for Karachi, Pakistan. Renew Energy 1:115–118

Akinoğlu BG, Ecevit A (1990) Construction of a quadratic model using modified Angström coefficients to estimate global solar radiation. Solar Energy 45:85–92

Angström A (1924) Solar terrestrial radiation. Q J R Meteorol Soc 50:121–126

Angström A (1929) On the atmospheric transmission of sun radiation and dust in the air. Geografiska Annaler 2:156–166

Angström A (1956) On the computation of global radiation from records of sunshine. Arkiv Geof 2:471–479

Angus RC (1995) Illuminance models for the United Kingdom. PhD thesis, Napier University, Edinburgh

Balling RC Jr, Cerveny RS (1983) Spatial and temporal variations in long-term normal percent possible radiation levels in the United States. J Clim Appl Meteorol 22:1726–1732

Baojun L, Dong W, Zhou M, Xu H (1995) Influences of optical fiber bend on solar energy optical fiber lighting. The second international conference on new energy systems and conversions, İstanbul, p 41

Barbaro S, Coppolino S, Leone C, Sinagra E (1978) Global solar radiation in Italy. Solar Energy 20:431

Beckman WA, Klein SA, Duffie JA (1977) Solar heating design by the f-chart model. Wiley-Interscience, New York

Benjamin JR, Cornell CA (1970) Probability, statistics and decision for civil engineers. McGraw Hill, New York, p 684

Bucciarelli LL (1984) Estimating the loss-of-power probabilities of stand alone photovoltaic solar energy systems. Solar Energy 32:205–209

Bucciarelli LL (1986) The effect of day-to-day correlation in solar radiation on the probability of loss of power in stand-alone photovoltaic solar energy systems. Solar Energy 36:11–14

Coulson KL (1975) Solar and terrestrial radiation. Academic, New York

Cressie NAL (1993) Statistics for spatial data. Wiley, New York

Davies JA, McKay DC (1989) Evaluation of selected models for estimating solar radiation on horizontal surfaces. Solar Energy 43:153–168

Davis J (1986) Statistic and data analysis in geology. Wiley, New York

Drummond AJ (1965) Techniques for the measurement of solar and terrestrial radiation fluxes in plant biological research: a review with special reference to arid zones. Proc Montpiller Symp, UNESCO

Gopinathan KK (1986) A general formula for computing the coefficients of the correlation connecting global solar radiation to sunshine duration. Solar Energy 41(6):499–502

Gordon JM, Reddy TA (1988) Time series analysis of daily horizontal solar radiation. Solar Energy 41:215–226

Gueymard C (1993) Mathematically integrable parametrization of clear-sky beam and global irradiances and its use in daily irradiation applications. Solar Energy 50:385–389

Gueymard CA (2003) The sun's total and spectral irradiance for solar energy applications and solar radiation models. Solar Energy 76:423–453

Gueymard C, Jidra P, Eatrada-Cajigai V (1995) A critical look at recent interpretations of the Angstrom approach and its future in global solar irradiation prediction. Solar Energy 54:357–363

Hay JE (1979) Calculation of monthly mean solar radiation for horizontal and inclined surfaces. Solar Energy 23:301–307

Hinrichsen K (1994) The Angström formula with coefficients having a physical meaning. Solar Energy 52:491–495
Hoyt DV (1978) Percent of possible sunshine and total cloud cover. Mont Weather Rev 105:648–652
Iqbal M (1979) Correlation of average diffuse and beam radiation with hours of bright sunshine. Solar Energy 23:169–173
Johnston RJ (1980) In: Multivariate statistical analysis in geography. Longman House, Essex, p 280
Journel AG, Huijbregts CJ (1978) In: Mining geostatistics. Academic, New York, p 600
Kimball HH (1919) Variations in the total and luminous solar radiation with geographical position in the United States. Mont Weather Rev 47:769–793
Kreider JF, Kreith F (1981) Solar energy handbook. McGraw-Hill, New York
Liu B, Jordan RC (1960) The interrelationship and characteristic distribution of direct, diffuse and total solar radiation. Solar Energy 4:1–4
Löf GOG, Duffie JA, Smith CO (1966) World distribution of solar radiation. Solar Energy 10:27–37
Ma CCY, Iqbal M (1984) Statistical comparison of solar radiation correlations. Monthly average global and diffuse radiation on horizontal surfaces. Solar Energy 33:143–148
Martinez-Lozano JA, Tena F, Onrubai JE, de la Rubia J (1984) The historical evolution of the Angström formula and its modifications: review and bibliography. Agric Forest Meteorol 33:109–128
Observers' Handbook (1969) HMSO, London
Page JK (1961) The estimation of monthly mean values of daily total short wave radiation on vertical and inclined surfaces from sunshine records for latitudes 40°N–40°S. In: Proceedings of United Nations conference on new resources of energy, Paper S/98, 4:378–380
Painter HE (1981) The performance of a Campbell–Stokes sunshine recorder compared with a simultaneous record of the normal incidence irradiance. Meteorol Mag 110:102–187
Palz W, Greif J (eds) (1996) European solar radiation atlas: solar radiation on horizontal and inclined surfaces, 3rd edn. Springer, Berlin
Perez R, Ineichen P, Seals R (1990) Modeling daylight availability and irradiance components from direct and global irradiance. Solar Energy 44:271
Prescott JA (1940) Evaporation from water surface in relation to solar radiation. Trans R Soc Austr 40:114–118
Rawlins F (1984) The accuracy of estimates of daily global irradiation from sunshine records for the United Kingdom. Meteorol Mag 113:187
Rietveld MR (1978) A new method for estimating the regression coefficients in the formula relating solar radiation to sunshine. Agric Meteorol 19:243–252
Şahin AD, Şen Z (1998) Statistical analysis of the Angström formula coefficients and application for Turkey. Solar Energy 62:29–38
Şahin AD, Kadioglu M, Şen Z (2001) Monthly clearness index values of Turkey by harmonic analysis approach. Energy Convers Manage 42:933–940
Şen Z (2001a) Mühendislikte matematik modelleme yöntemleri (Mathematical modeling principles in engineering) (in Turkish). Su Vakfı, İstanbul
Şen Z (2001b) Angström equation parameter estimation by unrestricted method. Solar Energy 71:95–107
Şen Z, Öztopal A (2003) Terrestrial irradiation: sunshine duration clustering and prediction. Energy Convers Manage 44:2159–2174
Şen Z, Şahin AD (2001) Spatial interpolation and estimation of solar irradiation by cumulative semivariograms. Solar Energy 71:11–21
Suehrcke H (2000) On the relationship between duration of sunshine and solar radiation on the earth's surface: Angström's equation revisited. Solar Energy 68:417–425
Suehrcke H, McCormick PG (1992) A performance prediction method for solar energy systems. Solar Energy 48:169–175

Thevenard D, Leng G, Martel S (2000) The RET screen model for assessing potential PV projects. In: Proceedings of the 28th IEEE photovoltaic specialists conference, 15–22 Sept, Anchorage, pp 1626–1629

Wahab AM (1993) New approach to estimate Angström coefficients. Solar Energy 51:241–245

WMO, World Meteorological Organization (1962) Commission for instruments and methods of observation. Abridged final report of the third session. WMO-No. 363, Geneva

WMO, World Meteorological Organization (2003) Manual on the global observing system. WMO-No. 544, Geneva

Chapter 5
Non-Linear Solar Energy Models

5.1 General

As already explained in the previous chapter different versions of linear AMs are in use extensively in solar energy studies for estimation of the global terrestrial solar radiation amounts from the sunshine duration data. However, atmospheric *turbidity* and *transmissivity*, planetary boundary layer turbulence, cloud thickness, and temporal and spatial variations cause embedding of non-linear elements in the solar radiation phenomena. Hence, the use of simple linear models cannot be justified physically except statistically without thinking about obtaining the model parameter estimations. In the literature, most often the linear models are either modified with the addition of extra terms in the hope of explaining the non-linear features or adjustment of the linear model parameters by relating them to geographical, meteorological, and other variables. Non-linearity in solar radiation and sunshine duration relationships is represented initially through classic statistical techniques by the addition of non-linear terms to the basic AM of Chap. 4. Some other researchers have incorporated the non-linear behavior in the models by expressing the linear model parameters in terms of each other or in terms of sunshine and solar radiation variables. Such models reduce to the classic linear solar radiation models under a set of specific assumptions.

In this chapter after the general review of available classic non-linear models, additional innovative non-linear models are presented with fundamental differences and distinctions. Fuzzy logic and genetic algorithm approaches are presented for the non-linear modeling of solar radiation from sunshine duration data.

5.2 Classic Non-Linear Models

Most of the sunshine duration based solar radiation estimation models are modifications of the AM (Chap. 4, Eq. 4.30). Some authors have suggested changing the

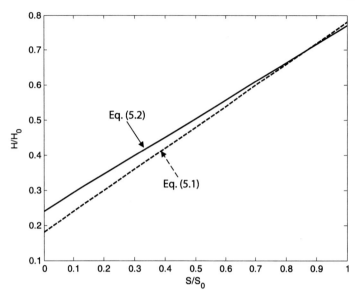

Fig. 5.1 Benson *et al.* (1984) seasonal model

model parameters a and b seasonally thus arriving at better estimations (Abdalla and Baghdady 1985; Benson *et al.*, 1984; Rietveld 1978). Barbara *et al.* (1978), Şen and Öztopal (2003), and Suleiman (1985) have expressed the global irradiation in terms of the sunshine duration and the geographic location. Hay (1984) related clouds and the atmospheric conditions to the solar radiation estimation. He proposed the use of the AM with a modified day-length instead of $\overline{S}$ and solar radiation that first hits the ground instead of $\overline{H}$ (Chap. 4). In the search for the non-linearity effects, Benson *et al.* (1984) have suggested modification of the AM with two six-month seasons, similar to the linear cluster model (LCM) as proposed by Şen and Öztopal (2003), namely, October–March and April–September periods leading to two different linear models as follows:

$$\frac{\overline{H}}{H_0} = 0.18 + 0.60 \frac{\overline{S}}{S_0} \tag{5.1}$$

and

$$\frac{\overline{H}}{H_0} = 0.24 + 0.54 \frac{\overline{S}}{S_0}, \tag{5.2}$$

respectively. It is to be noticed that although the summations of $a+b$ in these two models are the same, a and b have different values in the two periods. These linear models represent the scatter region by two non-parallel AMs as in Fig. 5.1.

Gopinathan (1988) has related the AM parameters to geographic elevation, h, and the ratio of sunshine duration as follows:

5.2 Classic Non-Linear Models

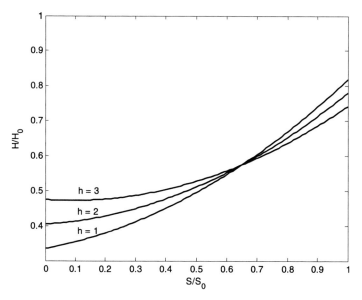

Fig. 5.2 Gopinathan (1988) model

$$a = 0.265 + 0.07h - 0.135\frac{\overline{S}}{S_0} \tag{5.3}$$

and

$$b = 0.265 - 0.07h - 0.325\frac{\overline{S}}{S_0}. \tag{5.4}$$

Substitution of these parameters into the AM (Eq. 4.30) leads to non-linear global solar radiation estimation models. The non-linearities are sought indirectly through the parameters' relationship to the sunshine duration ratio. For a set of elevation values this model is shown in Fig. 5.2. In general these models have concave shapes.

Ögelman et al. (1984) suggested adding a non-linear term to the AM, thus making the following *quadratic model*:

$$\frac{\overline{H}}{H_0} = a + b\frac{\overline{S}}{S_0} + c\left(\frac{\overline{S}}{S_0}\right)^2 \tag{5.5}$$

Based on this quadratic equation, Akınoğlu and Ecevit (1990) showed its superiority over other models in terms of global applicability. They applied this model to some Turkish data and finally obtained a suitable model that was better than the AM:

$$\frac{\overline{H}}{H_0} = 0.195 + 0.676\frac{\overline{S}}{S_0} - 0.142\left(\frac{\overline{S}}{S_0}\right)^2 \tag{5.6}$$

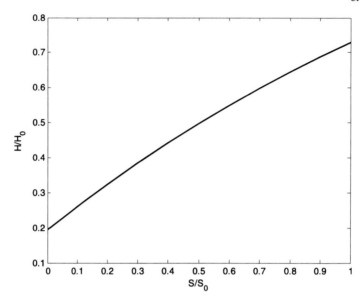

Fig. 5.3 Akınoğlu and Ecevit (1990) quadratic model

Figure 5.3 shows the graphical representation of this quadratic model. Compared to the Gopinathan model in Fig. 5.2, it has a convex curvature.

Higher-order polynomial non-linear models are also proposed in the solar energy literature and, in particular, Zabara (1986) correlated the AM parameters to the third power of the sunshine duration ratio as

$$a = 0.395 - 1.247 \frac{\overline{S}}{S_0} + 2.680 \left(\frac{\overline{S}}{S_0}\right)^2 - 1.674 \left(\frac{\overline{S}}{S_0}\right)^3 \tag{5.7}$$

and

$$b = 0.395 + 1.384 \frac{\overline{S}}{S_0} - 3.248 \left(\frac{\overline{S}}{S_0}\right)^2 + 2.055 \left(\frac{\overline{S}}{S_0}\right)^3. \tag{5.8}$$

After the substitution of these parameters into the AM, the Zabara model for Greece appears as in Figure 5.4, which has convex and concave parts in the overall performance.

Akınoğlu and Ecevit (1990) found a global relationship between the AM parameters by using the published a and b values for 100 locations all over the world and the relationship is suggested in the following quadratic form:

$$a = 0.783 - 1.509b + 0.892b^2. \tag{5.9}$$

5.2 Classic Non-Linear Models

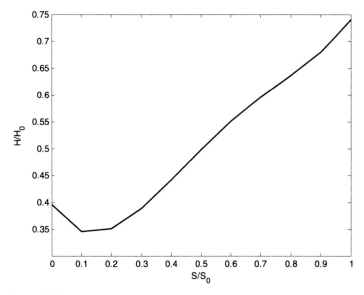

Fig. 5.4 Zabara (1986) model

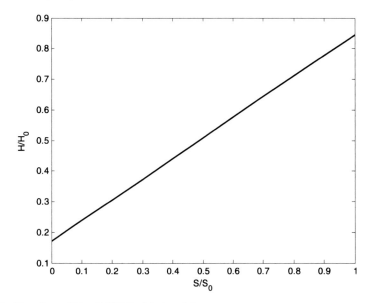

Fig. 5.5 Akınoğlu and Ecevit (1990) global model

Given any b value this expression provides a value and its substitution into the AM leads to the model appearance as in Fig. 5.5.

5.3 Simple Power Model (SPM)

Generally, the PDF of global solar radiation and sunshine duration data are unimodal but not necessarily Gaussian (Caughey et al., 1982; Chap. 4). Intense cloud areas indicate increase in the *turbulence* at the cloud top, which causes a decrease in the global solar radiation. A number of theoretical studies have shown the sensitivity of cloud irradiative properties to their spatial structure affecting the sunshine duration and subsequently the global irradiation amounts (Davies 1978; Joseph and Cahalan 1990). It is, therefore, significant to preserve in any estimation procedure the third- or higher-order statistical moments (Wahab 1993). One way of incorporating these moments in the model is the inclusion of a third power of the sunshine duration variable as

$$\frac{H}{H_0} = a + b\frac{S}{S_0} + c\left(\frac{S}{S_0}\right)^2 + d\left(\frac{S}{S_0}\right)^3, \qquad (5.10)$$

where c and d are additional model parameters. However, there are implied mathematical and physical assumptions in such *polynomial models* as follows:

1. In any modeling approach, parsimony (*i.e.*, to obtain the best result with the least number of parameters) is one of the requirements (Box and Jenkins 1970).
2. It is difficult to explain on physical grounds why a polynomial expression is adopted for modeling purposes apart from the mathematical convenience only. It is important to note that Eq. 5.10 includes integer powers (0, 1, 2, and 3) of the variable S/S_0.
3. In statistical literature, second-order statistics (variance) subsume first-order statistics (average), and third-order statistics (skewness) include first- and second-order statistics (Benjamin and Cornell 1970). In general, a polynomial model leads to imbedded redundancy in the model.

In order to avoid the aforementioned assumptions, a *simple power model* (SPM) with non-integer (decimal, fractional) power, p, is proposed (Şen 2007):

$$\frac{H}{H_0} = a_p + b_p\left(\frac{S}{S_0}\right)^p, \qquad (5.11)$$

where a_p, b_p, and c are the model parameters. Hence, an objective SPM structure is proposed by including a power on the sunshine duration term in the basic AM. The power may assume fractional values and if its value is equal to 1 then the model reduces to the AM approximation. If $p \neq 1$ then different types of convex and concave non-linear models are obtainable (Fig. 5.6).

Although the linear equation passes through the center $\overline{(H/H_0)}$ versus $\overline{(S/S_0)}$ of the scatter diagram, the SPM does not pass from this point. In contrast to the available linear and non-linear models in the literature, SPM subsumes more physical conditions. It is known that a_p corresponds to the relative diffuse radiation on any overcast day and the summation of a_p and b_p represents the relative cloudless-sky

5.3 Simple Power Model (SPM)

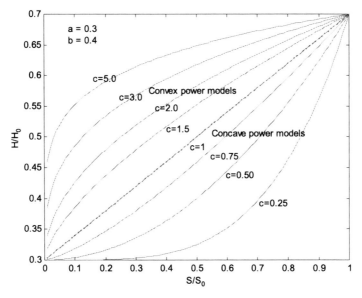

Fig. 5.6 Different SPMs

global irradiation. Hence, parameters a_p and b_p are reflections of extreme states, which correspond to full overcast and completely cloudless cases. The power p in Eq. 5.11, however, provides the whole expected dynamics from the SPM. Any non-linear effects due to the atmospheric composition and the joint behaviors of irradiation and sunshine duration phenomena are reflected in the power parameter.

Consideration of averages of solar radiation $\overline{(H/H_0)}$ and sunshine duration $\overline{(S/S_0)}$ renders Eq. 5.11, after some algebra, into

$$\overline{\frac{H}{H_0}} = a_p + b_p \left(\overline{\frac{S}{S_0}}\right)^{1/p}. \tag{5.12}$$

It is suggested that the SPM should be used in different parts of the world and its possible universal acceptance should be tested.

5.3.1 Estimation of Model Parameters

Since Eq. 5.12 has a non-linear form, the parameter estimations will be achieved through a non-linear least squares analysis. The basis of this methodology is to minimize the estimation error sum of squares (SS). This implies mathematically that the partial derivatives of the SS expression with respect to each model parameter should be set equal to zero. Detailed formulation of the non-linear least squares analysis is presented in Appendix B, which yields the following model parameter estimations:

$$a_p = \frac{1}{n}\sum_{i=1}^{n}\overline{\left(\frac{H}{H_0}\right)_i} - b_p \frac{1}{n}\sum_{i=1}^{n}\overline{\left(\frac{S}{S_0}\right)_i^{\frac{1}{p}}} \qquad (5.13)$$

and

$$b_p = \frac{\frac{1}{n}\sum_{i=1}^{n}\overline{\left(\frac{H}{H_0}\right)_i \left(\frac{S}{S_0}\right)_i^{\frac{1}{p}}} - \left[\frac{1}{n}\sum_{i=1}^{n}\overline{\left(\frac{H}{H_0}\right)_i}\right]\left[\frac{1}{n}\sum_{i=1}^{n}\overline{\left(\frac{S}{S_0}\right)_i^{\frac{1}{p}}}\right]}{\left[\frac{1}{n}\sum_{i=1}^{n}\overline{\left(\frac{H}{H_0}\right)_i^{\frac{1}{p}}}\right] - \left[\frac{1}{n}\sum_{i=1}^{n}\overline{\left(\frac{H}{H_0}\right)_i^{\frac{1}{p}}}\right]^2} . \qquad (5.14)$$

For $p = 1$ these two equations reduce to the classic regression line parameter estimates as in Eq. 4.31 and 4.32 of Chap. 4. Simultaneous analytical solution of the three parameters is not possible and, therefore, a numerical optimization procedure is used with the following steps (Şen 2007):

1. Initially, a small p value is chosen and with the solar radiation and sunshine duration data at hand, corresponding a_p and b_p parameter estimates are calculated from Eq. 5.13 and 5.14, respectively.
2. The SS error value is calculated from Eq. 5.12 by using Eq. B.2 from App. B:

$$SS = \sum_{i=1}^{n}\left[\overline{\left(\frac{H}{H_0}\right)} - a_p - b_p \overline{\left(\frac{S}{S_0}\right)_i^{\frac{1}{p}}}\right]. \qquad (5.15)$$

3. The slope, s_{j+1}, value of SS variation with p is calculated by taking the difference between the two successive SS values divided by the increment Δp in p as

$$s_{j+1} = \frac{(SS)_{j+1} - (SS)_j}{\Delta p}. \qquad (5.16)$$

Here, the counter $j+1$ shows the current but j the previous step calculations in the numerical optimization procedure.

4. If s_{j+1} are less than a pre-selected critical value, say, s_{cr}, then the optimization procedure is stopped with model parameter estimates. In practice, most often $\pm 5\%$ error band is admissible which implies that $s_{cr} = 0.05$.

   Steps 2-4 are repeated until the optimum estimation point is reached according to the $s_{j+1} \leq s_{cr}$ criterion. For the implementation of the methodology eight solar radiation measurement station records are considered from different climatic zones in Turkey (see Fig. 4.6 and Table 4.3).

   In order to obtain the optimum parameter estimates, the initial p value is assumed as 0.01, the p increment as $\Delta p = 0.01$, and, finally, the critical value as $s_{cr} = 0.05$. The parameter estimations from the non-linear least squares pro-

Table 5.1 SP model and AM estimations

Station name	SP				AM		
	a_p	b_p	p	SS	a	b	SS
Adiyaman	0.228	0.285	1.550	0.130	0.302	0.215	0.179
Afyon	0.429	0.256	0.760	0.160	0.395	0.277	0.165
Antalya	0.198	0.511	1.530	0.375	0.332	0.282	0.387
Diyarbakır	0.193	0.535	1.150	0.376	0.332	0.382	0.387
İstanbul	0.166	0.448	1.940	0.215	0.295	0.355	0.225
Izmir	0.273	0.382	1.210	0.355	0.317	0.310	0.370
Kastamonu	0.350	0.223	0.640	0.155	0.317	0.227	0.161
Konya	0.409	0.247	0.800	0.501	0.381	0.266	0.511

cedure are given in Table 5.1 with AM parameter estimates. The use of these parameters with the sunshine duration data at any site yields the solar irradiance estimates according to the non-linear model Eq. 5.11.

The use of the AM and the SPM does not make significant differences at the extreme weather conditions such as overcast and clear-sky situations, but most of the time, especially in subtropical regions, the weather situation occurs more often in between. The real divergence between the AM and the SPM appears at the intermediate sunshine duration and solar radiation data values. One can conclude physically that toward the equator in the tropical and polar regions the use of the AM is more reliable than in the subtropical regions. Physically, this implies that the power, p, comes close to 1 toward the polar and equator regions. The deviations from 1 are more in the subtropical regions. In the desert regions of the world, the AM approach is more reliable, but in the mountainous regions, with frequent partially cloudy weather, use of the SPM is preferable for more reliable predictions of the global solar radiation values from the measured sunshine duration data.

5.4 Comparison of Different Models

In order to compare all the models, they are plotted with the same a and b values on the same Cartesian coordinate system. For the appreciation of parameter changes (a and b) on the models, Fig. 5.7 is prepared for four different cases as follows: ($a = 0.1$; $b = 0.75$), ($a = 0.15$; $b = 0.85$), ($a = 0.15$; $b = 0.80$), and ($a = 0.15$; $b = 0.70$). In the figure, p is considered as a set of values that leads to a set of SPM curves that cover a significant part of the Cartesian coordinate system.

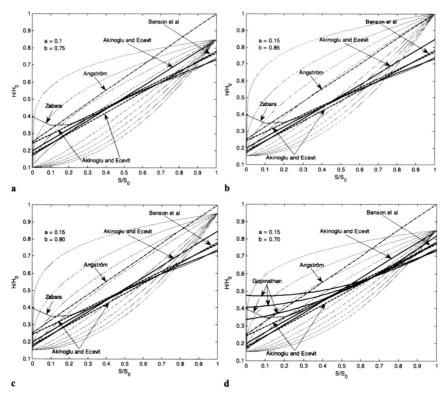

Fig. 5.7 a–d. All models together **a** for $a = 0.1$ and $b = 0.75$, **b** for $a = 0.15$ and $b = 0.85$, **c** for $a = 0.15$ and $b = 0.80$, and **d** for $a = 0.15$ and $b = 0.70$

5.5 Solar Irradiance Polygon Model (SIPM)

Classic approaches expressing the solar global irradiation in terms of sunshine durations are abundant in the literature and they include mostly linear and to a lesser extent non-linear relationships between these variables. None of them provide within-year (monthly) variations in the parameters. Herein, a *solar irradiance polygon model* (SIPM) is presented for evaluating qualitatively and quantitatively the within-year variations in the solar energy variables. On the basis of the SIPM, monthly, seasonal, and annual parameters of the classic AM can be calculated. The meteorological variability reflects itself in the astronomical extraterrestrial irradiation H_0 and sunshine duration S_0, *i. e.*, length of the day values, in two ways:

1. The astronomical *extraterrestrial irradiation* and sunshine duration are shortened due to meteorological and atmospheric events which are measured at a solar station as meteorological solar radiation H and sunshine duration S, such that $S < S_0$ and $H < H_0$.

5.5 Solar Irradiance Polygon Model (SIPM)

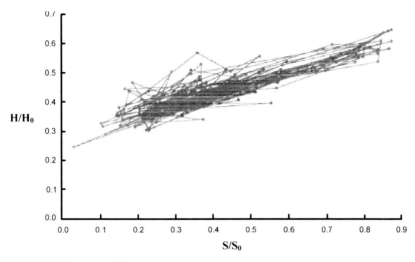

Fig. 5.8 Daily data scatter diagram

2. The shortening effect is not definite but might be in the form of different and random amounts during a day or a month depending on the climate and weather conditions.

As usual a Cartesian coordinate system is used for the scatter of points and then linear or non-linear expressions are suggested and subsequently by the least squares technique the model parameters are estimated (Chap. 4, Sect. 4.4.1). By considering the time sequence of points on the scatter diagram, successive straight lines between two observations can be traced. Such an application leads to a very complex and chaotic pattern for daily data and hence the pattern cannot be appreciated as shown in Fig. 5.8.

No doubt, there should be a certain pattern due to at least the astronomical effects such as month to month or season to season periodic effects. When monthly or seasonal averages of the same data set are considered, the pattern becomes visually clear and interpretable as in Fig. 5.9

A close inspection indicates that there emerges a polygon with 12 sides and vertices in a monthly sequential order which is referred to as the SIPM, where polygons depict the non-linearity, if any. Scatter in Fig. 5.9 is comparatively smaller than Fig. 5.8. Monthly, SIPMs at any station are similar to Fig. 5.9 with different features of width, peripheral length, side lengths, and areal extent depending on the geomorphologic characteristic of the station site, its altitude, longitude, and especially weather conditions. Hence, apart from the scatter of points, SIPMs provide a monthly variation sequence. Since, it is known physically that surface global solar radiation is positively related to the sunshine data, all the SIPMs exhibit high (low) values of extraterrestrial solar radiation following high (low) values of the sunshine data. In general, these diagrams provide the following benefits over the classic models (Şen and Şahin 2000):

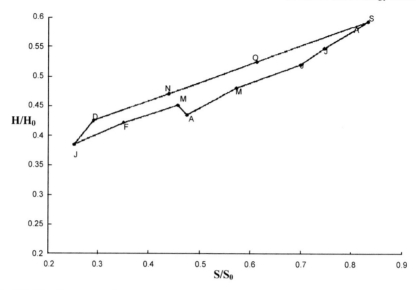

Fig. 5.9 Monthly average data scatter diagram

1. They are closed polygons, which indicate that the global solar radiation and sunshine duration processes evolve periodically within a year. However, on the top of such a periodicity, there are also the local meteorological effects. The reason for different SIPMs at different sites is due to differences in the weather and climate conditions, in addition to longitude, latitude, and altitude values.
2. Each side shows transition, *i.e.*, variation of the solar global irradiation amount with the sunshine duration between two successive months.
3. Similar to the regression straight line concept where the slope is related to parameter b, it is possible to calculate the slope between the two successive months, say i and $i-1$ as

$$b_i = \frac{\left(\frac{H}{H_0}\right)_i - \left(\frac{H}{H_0}\right)_{i-1}}{\left(\frac{S}{S_0}\right)_i - \left(\frac{S}{S_0}\right)_{i-1}} \quad (i = 2, 3, \ldots, 12). \quad (5.17)$$

Herein, it is referred to as the Monthly Global Solar Irradiation Change (MGSIC) with the sunshine duration. Şahin and Şen (1998) have employed this equation in their study for the statistical analysis of the AM parameter assessments (Eq. 4.34). In fact, it is equivalent to the derivation of the global solar irradiance with respect to the sunshine duration (Eq. 4.33). The smaller the time interval, the closer is this ratio to the mentioned derivation. It is to be noticed that such an interpretation cannot be attached to the b parameter estimation in the AM, which is based on the classic regression equation.
4. Another piece of information that can be deduced from SIPMs is the direction of the H/H_0 versus S/S_0 variation. The polygons are closed and hence there are

5.5 Solar Irradiance Polygon Model (SIPM)

two possible revolution directions, either clockwise or anti-clockwise. Hence, in some monthly durations the MGSIC values become negative in the sense that the global solar radiation and the sunshine duration both start to decrease with time. This is contrary to what can be deduced from the AM coefficient, b, which does not provide any information about the direction of the change.

5. The lengths of polygon sides give information on weather and astronomical change from one month to another. For instance, short lengths show that the changes are not significant. This is especially true if the weather conditions have remained almost the same during the transition between two successive months. For instance, one type of clear, hazy, overcast, partly cloudy, or cloudy sky conditions during such a transition causes these lengths to be short.
6. Comparison of two successive sides also provides information about the change of solar radiation rate from one month to other. If the angle between the two sides is negligibly small it is then possible to infer that the weather conditions have remained rather uniform.
7. The more the contribution of *diffuse solar radiation* on the global irradiation the wider will be the SIPM.
8. Depending to the closeness of each side to the vertical or horizontal directions there are different interpretations. For instance, in the case of a nearly horizontal side, there is no change in the global solar irradiance, which means that the effect of the weather has been such that it remained almost stable.
9. Each polygon has rising and falling limbs and, hence, shows two complementary periodic cycles. However, the number of months in each limb might not be equal to each other, depending on the meteorological effects and the station location.
10. The SIPMs provide two values for a given constant H/H_0 (or S/S_0) each of which lies on a different limb as referred to in the previous step. Hence, the difference between these two values yields the domain of change for the given constant variable value.
11. For comparison purposes one can plot two or more SIPMs according to latitudes, longitudes, or altitudes.

Although the application of the SIP concept is accomplished for all 29 stations in Table 4.3, the results are presented only for the İstanbul and Ankara stations in Turkey. Since Turkey lies in the northern hemisphere most often the SIPMs will have a clockwise rotation throughout year, however, depending on the weather conditions some polygons might have two loops. The SIPMs are presented in Fig. 5.10 for average monthly data inform the İstanbul and Ankara stations.

In general, all the SIPMs can be grouped into two categories, those with two loops (Fig. 5.10a) and those with a single (Fig. 5.10b) loop. Their interpretations must consider the geomorphologic features of the vicinity of the station location and the weather conditions in addition to the lower atmospheric stability. It is, therefore, necessary to talk about the common features of these SIPMs as follows:

1. All the SIPMs have high H/H_0 values for high S/S_0 values, and therefore, solar radiation increases with an increase in sunshine duration.

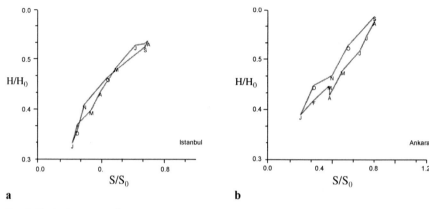

Fig. 5.10 a,b. SIPM for **a** İstanbul and **b** Ankara

2. SIPMs provide a basis for the monthly or seasonal global solar radiation-sunshine duration variability within one year. For instance, the least amounts occur in January or February but they attain their maximum in August or September. These periods are very close to *solstices*. In most of the SIPMs, the polygons cross to another loop near the spring *equinox* which is not the case during the summer solstice. This is tantamount to saying that with the start of warming during the spring season the summer equinox does not show any transitional affect as a node at the intersection of two loops.
3. The rising limbs of SIPMs start from January and continue until September then a falling limb starts and completes the yearly period.
4. The slopes of each side in SIPMs show the rate of global solar radiation change with sunshine duration between two successive months. The model parameters (a_s and b_s) for each side of the SIPM polygon are presented in Table 5.2. Herein, a_s and b_s are the intercept of each polygon side on the vertical axis (H/H_0) corresponding to S/S_0 and b_s is the slope of the same side.

It is obvious that a_s and b_s values vary from month to month during one year. Compared with the AM coefficients these show great differences in value. However, the monthly average throughout year became more stable and close to the AM as shown in Table 5.3.

A close inspection of Table 5.3 shows that there are negative a_s and b_s values especially in monthly and to a lesser extent in seasonal durations, but for the annual case they are all positive. The negative b_s values indicate that during this month increases in the sunshine duration have caused decrease in the global solar radiation. Furthermore, short sunshine duration gives rise to high solar radiation. This is physically possible if a hot air mass develops over the measurement station area due to some meteorological features such as wind speed. It may also be due to the *urban heat island* effect for measurement stations close to city centers. For such physical interpretations one should be confident that there are no measurement errors in the solar energy variable records. The SIPM approach provides a basis for the calculation of seasonal a_s and b_s values as

5.5 Solar Irradiance Polygon Model (SIPM)

Table 5.2 a_s and b_s coefficients from SIPM

Month	Station number 1	2	3	4	5	6	7	8	9	10	11	12	13	14	15
	a	a	a	a	a	a	a	a	a	a	a	a	a	a	a
J–F	−0.03	0.53	0.30	0.06	−0.27	0.33	0.27	0.00	−0.10	0.20	0.12	0.31	−0.19	0.23	0.46
F–M	−0.50	0.28	0.39	0.32	0.34	0.31	−0.17	0.33	0.24	0.36	0.07	0.40	−0.01	0.33	0.48
M–A	0.25	0.27	0.60	0.28	−0.39	0.29	0.36	0.33	0.18	0.30	0.29	−0.64	0.25	0.41	−0.25
A–M	0.41	0.40	0.44	0.32	−0.49	0.20	0.44	0.15	0.23	0.26	0.29	0.32	0.27	0.30	0.35
M–J	0.35	0.29	0.33	0.27	0.19	0.30	0.22	0.35	0.23	0.22	0.39	0.36	0.25	0.20	0.35
J–J	0.00	0.39	0.31	0.31	0.21	0.10	0.50	0.09	0.07	0.30	0.20	0.29	0.78	0.24	0.39
J–A	0.55	0.49	0.13	0.31	0.93	0.18	2.53	0.11	0.17	−0.21	0.13	0.13	0.55	0.56	0.40
A–S	0.46	1.12	0.70	0.40	−0.09	0.26	2.42	0.43	0.59	1.03	0.70	1.57	1.22	0.40	0.38
S–O	0.27	0.38	0.34	0.23	−0.52	0.35	0.20	0.45	0.25	0.35	0.30	0.41	0.39	0.26	0.40
O–N	0.05	0.29	0.32	0.29	0.11	0.23	0.29	0.31	0.15	0.18	0.31	0.32	0.16	0.22	0.43
N–D	−0.69	0.22	0.40	0.18	0.10	0.40	−0.38	0.08	0.25	0.64	0.16	0.37	−0.25	0.35	0.41
D–J	0.07	−0.52	0.42	0.50	−0.10	0.24	0.00	1.25	−0.12	0.59	0.38	0.34	0.06	0.30	0.64
Spring	0.34	0.32	0.46	0.29	−0.23	0.26	0.34	0.28	0.21	0.26	0.32	0.01	0.26	0.30	0.15
Summer	0.34	0.66	0.38	0.34	0.35	0.18	1.82	0.21	0.27	0.37	0.34	0.66	0.85	0.40	0.39
Autumn	−0.12	0.30	0.35	0.23	−0.10	0.33	0.04	0.28	0.22	0.39	0.26	0.36	0.10	0.28	0.42
Winter	−0.15	0.10	0.37	0.30	−0.01	0.29	0.03	0.53	0.01	0.39	0.19	0.35	−0.05	0.29	0.53
Annual	0.10	0.34	0.39	0.29	0.00	0.27	0.56	0.32	0.18	0.35	0.28	0.35	0.29	0.32	0.37
	b	b	b	b	b	b	b	b	b	b	b	b	b	b	b
J–F	1.00	−0.32	0.57	1.38	1.95	0.25	0.43	1.06	1.45	0.57	0.72	0.52	1.40	0.55	0.08
F–M	2.00	0.27	0.29	0.32	0.26	0.30	1.29	0.37	0.23	0.02	0.85	0.24	1.01	0.29	0.05
M–A	0.43	0.28	−0.18	0.43	2.17	0.33	0.28	0.36	0.42	0.20	0.35	2.63	0.49	0.11	1.61
A–M	0.14	0.06	0.15	0.32	2.42	0.50	0.15	0.67	0.31	0.31	0.36	0.36	0.45	0.32	0.28
M–J	0.25	0.22	0.36	0.43	0.75	0.31	0.48	0.36	0.30	0.37	0.20	0.27	0.47	0.48	0.27
J–J	0.75	0.10	0.38	0.37	0.70	0.60	0.10	0.09	0.54	0.26	0.45	0.38	−0.16	0.42	0.21
J–A	0.00	−0.01	0.62	0.37	−0.93	0.50	−2.47	0.67	0.41	1.00	0.54	0.62	0.11	0.05	0.21
A–S	0.12	−0.69	−0.09	0.25	1.38	0.40	−2.34	0.27	−0.16	−0.73	−0.14	−1.33	−0.65	0.24	0.22
S–O	0.38	0.12	0.38	0.50	2.35	0.30	0.54	0.25	0.32	0.24	0.37	0.28	0.32	0.40	0.22
O–N	0.75	0.25	0.41	0.37	0.88	0.50	0.41	0.46	0.52	0.57	0.34	0.43	0.64	0.47	0.17
N–D	2.00	0.39	0.22	0.77	0.90	0.14	1.63	0.92	0.21	−0.73	0.70	0.28	1.43	0.16	0.23
D–J	0.80	1.95	0.16	−0.67	1.45	0.60	0.94	−1.56	1.52	−0.60	0.14	0.40	0.76	0.32	−0.66
Spring	0.27	0.19	0.11	0.39	1.78	0.38	0.30	0.46	0.34	0.30	0.30	1.09	0.47	0.30	0.72
Summer	0.29	−0.20	0.30	0.33	0.30	0.50	−1.57	0.54	0.26	0.17	0.28	−0.11	−0.23	0.24	0.21
Autumn	1.04	0.25	0.34	0.55	1.38	0.31	0.86	0.54	0.35	0.02	0.47	0.33	0.80	0.34	0.21
Winter	1.27	0.63	0.34	0.34	1.22	0.38	0.88	−0.04	1.07	−0.01	0.57	0.39	1.06	0.39	−0.18
Annual	0.72	0.22	0.27	0.40	1.19	0.39	0.12	0.33	0.51	0.12	0.41	0.42	0.52	0.32	0.24

Table 5.2 (continued)

Month	Station number 16	17	18	19	20	21	22	23	24	25	26	27	28	29
	a	a	a	a	a	a	a	a	a	a	a	a	a	a
J–F	0.33	0.11	0.35	1.93	0.45	0.32	0.37	0.27	0.27	0.23	0.94	0.28	0.20	0.57
F–M	0.32	0.29	0.41	0.37	0.39	0.43	0.38	0.43	0.45	0.38	−0.14	0.43	−0.27	0.68
M–A	0.60	0.19	0.69	0.22	1.02	−0.10	0.51	0.78	0.57	0.15	0.41	0.28	0.44	0.61
A–M	0.39	0.25	0.37	0.34	0.43	0.30	0.27	0.38	0.34	0.37	0.49	0.18	0.25	0.53
M–J	0.40	0.30	0.26	0.29	0.49	0.32	0.30	0.35	0.25	0.26	0.12	0.26	0.39	0.37
J–J	0.20	0.50	0.34	0.44	0.16	0.15	0.15	0.45	0.30	0.29	0.38	0.38	0.39	0.62
J–A	0.30	0.33	0.25	0.39	0.42	0.25	0.71	0.46	0.35	0.44	0.39	0.42	0.60	1.10
A–S	0.93	0.00	0.59	0.34	1.55	0.72	0.88	1.23	0.71	0.63	0.20	0.47	0.28	0.27
S–O	0.42	0.35	0.37	0.36	0.39	0.26	0.31	0.44	0.45	0.25	0.38	0.29	0.33	0.50
O–N	0.38	0.30	0.34	0.34	0.42	0.36	0.32	0.40	0.36	0.30	0.27	0.13	0.18	0.38
N–D	0.34	0.09	0.33	−0.26	0.99	0.36	0.38	0.33	0.26	0.52	−1.12	0.50	0.33	0.86
D–J	0.25	0.18	0.44	−0.11	0.40	0.24	0.38	0.46	0.33	0.35	0.41	0.38	0.17	0.57
Spring	0.46	0.25	0.44	0.28	0.65	0.17	0.36	0.51	0.39	0.26	0.34	0.24	0.36	0.50
Summer	0.48	0.28	0.39	0.39	0.71	0.37	0.58	0.71	0.45	0.45	0.32	0.42	0.42	0.66
Autumn	0.38	0.25	0.34	0.15	0.60	0.33	0.34	0.39	0.35	0.36	−0.16	0.31	0.28	0.58
Winter	0.30	0.19	0.40	0.73	0.42	0.33	0.37	0.39	0.35	0.32	0.40	0.36	0.03	0.61
Annual	0.41	0.24	0.39	0.39	0.59	0.30	0.41	0.50	0.38	0.35	0.23	0.33	0.27	0.59
	b	b	b	b	b	b	b	b	b	b	b	b	b	b
J–F	0.43	1.04	0.18	−3.20	0.05	0.34	0.18	0.43	0.58	0.52	−0.88	0.54	0.45	0.02
F–M	0.47	0.32	0.06	0.21	0.24	−0.02	0.17	0.15	0.10	0.14	1.30	−0.13	1.70	−0.23
M–A	−0.21	0.60	−0.48	0.50	−1.28	1.27	−0.14	−0.50	−0.13	0.64	0.26	0.40	−0.13	−0.09
A–M	0.22	0.46	0.10	0.30	0.19	0.23	0.37	0.23	0.30	0.25	0.11	0.70	0.35	0.05
M–J	0.21	0.36	0.28	0.37	0.07	0.19	0.31	0.32	0.44	0.43	0.73	0.50	0.03	0.32
J–J	0.50	0.05	0.17	0.17	0.58	0.47	0.53	0.19	0.38	0.39	0.33	0.26	0.04	−0.02
J–A	0.36	0.29	0.29	0.24	0.22	0.32	−0.19	0.20	0.32	0.22	0.30	0.18	−0.48	−0.60
A–S	−0.41	0.77	−0.11	0.29	−1.29	−0.30	−0.40	−1.09	−0.11	0.00	0.56	0.09	0.33	0.45
S–O	0.26	0.25	0.17	0.31	0.32	0.37	0.34	0.28	0.18	0.46	0.33	0.45	0.22	0.18
O–N	0.34	0.36	0.22	0.34	0.27	0.16	0.33	0.28	0.33	0.39	0.49	0.87	0.56	0.37
N–D	0.45	1.10	0.23	1.51	−1.30	0.16	0.17	0.34	0.57	−0.14	2.94	−0.25	0.15	−0.54
D–J	0.32	0.73	−0.01	1.21	0.23	0.26	0.17	0.04	0.37	0.35	0.31	0.06	0.54	0.00
Spring	0.07	0.47	−0.03	0.39	−0.34	0.56	0.18	0.02	0.20	0.44	0.37	0.53	0.08	0.09
Summer	0.15	0.37	0.11	0.23	−0.16	0.16	−0.02	−0.23	0.20	0.20	0.40	0.18	−0.04	−0.05
Autumn	0.35	0.57	0.21	0.72	−0.24	0.23	0.28	0.30	0.36	0.24	1.25	0.36	0.31	0.00
Winter	0.57	0.69	0.07	−0.59	0.17	0.19	0.17	0.21	0.35	0.34	0.24	0.16	0.90	−0.07
Annual	0.29	0.53	0.09	0.19	−0.14	0.29	0.15	0.07	0.28	0.30	0.56	0.31	0.31	−0.01

Table 5.3 SIPM and classic approach a and b values with relative error

Station number	Classic		\multicolumn{3}{c}{SIP}			
	a	b	a_s			b_s		
			Min.	Ave.	Max.	Min.	Ave.	Max.
1	0.3	0.3	−0.7	0.1	0.6	0.0	0.7	2.0
2	0.3	0.2	−0.5	0.3	1.1	−0.7	0.2	1.9
3	0.4	0.3	0.1	0.4	0.7	−0.2	0.3	0.6
4	0.3	0.4	0.1	0.3	0.5	−0.7	0.4	1.4
5	0.4	0.2	−0.5	0.0	0.9	−0.9	1.2	2.4
6	0.3	0.3	0.1	0.3	0.4	0.1	0.4	0.6
7	0.3	0.4	−0.4	0.6	2.5	−2.5	0.1	1.6
8	0.3	0.4	0.0	0.3	1.3	−1.6	0.4	1.1
9	0.2	0.4	−0.1	0.2	0.6	−0.2	0.5	1.5
10	0.3	0.3	−0.2	0.4	1.0	−0.7	0.1	1.0
11	0.3	0.3	0.1	0.3	0.7	−0.1	0.4	0.9
12	0.4	0.3	−0.6	0.3	1.6	−1.3	0.4	2.6
13	0.2	0.5	−0.3	0.3	1.2	−0.6	0.5	1.4
14	0.3	0.3	0.2	0.3	0.6	0.0	0.3	0.6
15	0.4	0.2	−0.3	0.4	0.6	−0.7	0.2	1.6
16	0.4	0.3	0.2	0.4	0.9	−0.4	0.3	0.8
17	0.3	0.4	0.0	0.2	0.5	0.0	0.5	1.1
18	0.4	0.2	0.3	0.4	0.7	−0.5	0.1	0.3
19	0.3	0.3	−0.3	0.4	1.9	−3.2	0.2	1.5
20	0.5	0.1	0.2	0.6	1.5	−1.3	−0.1	0.6
21	0.3	0.2	−0.1	0.3	0.7	−0.3	0.3	1.3
22	0.4	0.2	0.1	0.4	0.9	−0.4	0.2	0.5
23	0.4	0.2	0.3	0.5	1.2	−1.1	0.1	0.4
24	0.4	0.3	0.3	0.4	0.7	−0.1	0.3	0.6
25	0.3	0.4	0.2	0.3	0.6	−0.1	0.3	0.6
26	0.3	0.4	−1.1	0.2	0.9	−0.9	0.6	2.9
27	0.3	0.3	0.1	0.3	0.5	−0.3	0.3	0.9
28	0.3	0.4	−0.3	0.3	0.6	−0.5	0.3	1.7
29	0.5	0.1	0.3	0.6	1.1	−0.6	0.0	0.5

averages of the 3 months of each season. On the other hand, it is obvious from Table 5.3 that there are significant differences between the average a_s and b_s values determined from SIPM and AM approaches. With the SIPM approach, it is possible to make internal estimates of minimum and maximum for a_s and b_s parameters.

5. It is obvious from the appearance of SIPMs that within a year there are two H/H_0 (S/S_0) values corresponding to a given constant S/S_0 (H/H_0) value. This point cannot be captured from the direct application of the AM.
6. In the great majority of the SIPMs the same sunshine duration value, S/S_0, gives higher global solar radiation during the autumn months than in spring. This is due to the fact that the terrain has already warmed up in the summer period and there is a delay prior to cooling. On the other hand, in the spring period although warming up has already started the effect of winter cooling still prevails. Hence,

5.5 Solar Irradiance Polygon Model (SIPM)

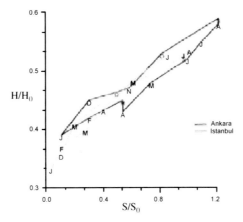

Fig. 5.11 SIPs at different altitudes

SIPMs show that solar energy production possibilities in the autumn months are more than in spring.

7. As the altitude of the station increases the corresponding SIPMs become wider. This means that for the same sunshine duration in different seasons the solar energy amounts become significantly different from each other. Another point is that these stations are very close to high mountains on the top of which the snow cover exists throughout the year with few exceptions. Hence, the climate in these regions is kept rather cool and dry during the year.

Further interpretations are possible by comparing the SIPMs at different locations, altitudes, latitudes, and longitudes. For instance, in Fig. 5.11 two important cites, namely, İstanbul along the coastal region on the Bosphorus (straight) connecting the Black Sea to the Marmara, Aegean, and Mediterranean Seas, and Ankara in central Anatolia with continental climate features, are shown just for comparison purposes.

Although both station's monthly values lie almost along straight lines that can be represented by the AM, the possible production of solar energy in Ankara is more than İstanbul at all months. Since, Ankara is located at an elevation almost 600 m more than İstanbul, its SIPM is wider. The changes of SIPM parameters month by month for İstanbul and Ankara stations are shown in Fig. 5.12.

It is obvious that $a_s + b_s$ values do not have very significant fluctuations about their arithmetic averages but a_s and b_s show opposite fluctuations to each other. This is to say that an increase (decrease) in a_s is associated with a decrease (increase) in b_s. However, their summations are more stable.

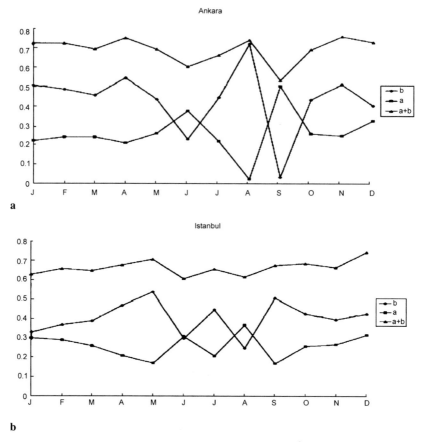

Fig. 5.12 a,b. Monthly a_s, b_s, and $a_s + b_s$ values for **a** Ankara and **b** İstanbul locations

5.6 Triple Solar Irradiation Model (TSIM)

Although there are multiple regression models that relate the solar radiation to variables such as sunshine duration, humidity, temperature, elevation, *etc.*, they are based on the restrictive assumptions (Sect. 4.4.1) as required by the regression technique methodology (Şen 2001; Şen and Şahin 2000). Such regression models do not exhibit variations in an elastic manner, but rather in a deterministic form. They provide a mathematical relationship, but, due to an increase in the number of variables, the error source might also increase and the model reliability becomes questionable. In order to avoid such restrictive situations, it is suggested in this section to draw contour lines of solar radiation values based not only on the sunshine duration, but additionally, on the *relative humidity* (RH). Hence, *triple solar irradiation model*s (TSIM) indicate without model restrictive affects the natural variability of solar radiation with sunshine duration and RH. Depending on the significance of the third

5.6 Triple Solar Irradiation Model (TSIM)

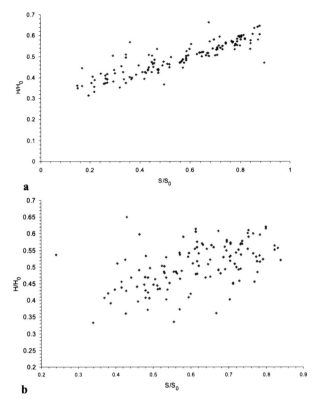

Fig. 5.13 a,b. Stations and some of their regional properties

variable, one can consider the TSIM approach with variables other than RH, such as temperature, precipitation, evaporation, *aerosol* concentration, *etc.*

Classic scatter diagrams are shown in Fig. 5.13, where in Fig. 5.13a the scatter is very restrictive but in Fig. 5.13b the scatter is comparatively wide indicating that it is preferable to consider another, *i. e.*, third variable in order to explain these scatters more physically.

Narrow scatter implies that solar radiation is affected and can be explained by sunshine duration to a great extent. Regression models are successful in cases of narrow scatter diagrams, but in very wider scatters it is preferable to import a third variable for better physical estimations. For this purpose, solar radiation is taken as the dependent variable with two independent variables, sunshine duration and RH. It is possible to construct a triple-regression equation between solar radiation, sunshine duration, and RH variables but such a model will suffer from the restrictive assumptions of the regression approach (Chap. 4).

In the TSIM estimation procedure, the vertical Cartesian axis is allocated for sunshine duration ratio and the horizontal axis is for the RH. The contour lines are drawn for the solar radiation ratios as dependent variable. The TSIM graphs show the following general points:

1. Solar radiation variation based on sunshine duration and RH variation.
2. Solar radiation variation with sunshine duration for any given level of RH.
3. Solar radiation variation with RH for any given level of sunshine duration value.
4. Solar radiation maxima occurrences at RH and sunshine duration values.
5. Solar radiation minima occurrences at sunshine duration and RH values.
6. Solar radiation variation for any combination of sunshine and RH values.
7. Locations of nearly clear weather conditions based on sunshine duration and RH values.
8. Locations of nearly overcast weather conditions based on sunshine duration and RH values.
9. Average and standard deviation values of solar radiation for given ranges of RH and sunshine duration, and hence it is possible to obtain the arithmetical average and standard deviation variations of solar radiation by sunshine duration and RH values. This is very helpful in deciding the error limitations concerning upper and lower boundaries in any solar radiation estimation.

Under the light of the previous step, one can obtain estimation of solar radiation for a given pair of sunshine duration and RH values. The TSIM graphs may be prepared for different time periods such as hourly, daily, weekly, and monthly. Comparison of two or more TSIM graphs at different locations helps to identify climatic differences to a significant extent.

The implementation of the TSIM is presented for Adana and Diyarbakır stations in Turkey (Fig. 4.6 and Table 4.3). Adana lies in the southern part with its typical Mediterranean climatic features where winters are relatively short with cool periods and summers have high temperatures and RH. Fig. 5.14a indicates that the solar radiation values change from 0.35 to 0.75. This figure is rather rough which indicates unstable climatic conditions. The maxima of solar radiation are for high sunshine duration and RH values. The minima take place at low sunshine durations but with RH confined between.

Diyarbakır region, on the other hand, is completely in the semi-arid region of Turkey in the southeastern part with dry, and hence, low RH but long periods of sunshine duration, especially, in the summer. Figure 5.14b indicates solar radiation contour changes from almost 0.30 to 0.65 with minima at low sunshine duration but high RH area. Maxima are for high sunshine duration but moderate RH values. It is possible to make solar radiation estimations from these maps for a given pair of sunshine duration and RH values. Although for any given sunshine duration there is only one solar radiation estimate from the AM method, in Table 5.4, solar radiation estimates are given at different RH levels.

5.6 Triple Solar Irradiation Model (TSIM)

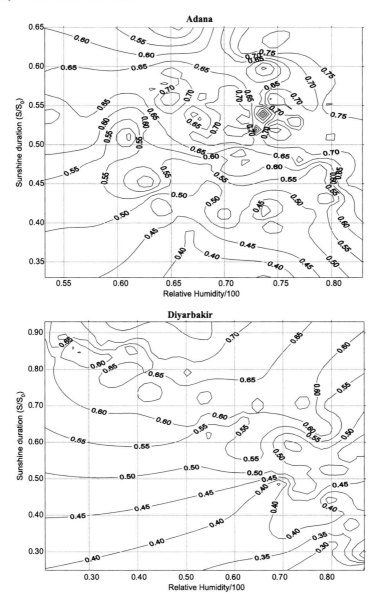

Fig. 5.14 a,b. TSI graph for **a** Adana and **b** Diyarbakır

The differences between the AM and TSIM estimations are within the practically acceptable ranges, but the TSIM graphs are more reliable because they take into consideration the contribution of the third variable which is the RH.

Table 5.4 TISM parameters

Measured S/S_0	H/H_0 estimation by AM	H/H_0 estimation by TSIM RH levels					
		0.40	0.50	0.55	0.60	0.70	0.80
ADANA							
0.20	0.37170	0.44	0.43		0.42	0.39	0.34
0.30	0.40535	0.45	0.45		0.45	0.42	0.46
0.40	0.43900	0.46	0.46		0.45	0.44	0.44
0.50	0.47265	0.47	0.46		0.45	0.47	0.48
0.60	0.50630	0.50	0.50		0.51	0.52	0.52
0.70	0.53995	0.54	0.53		0.57	0.57	0.55
0.80	0.57360	0.58	0.58		0.61	0.61	0.58
DIYARBAKIR							
0.35	0.43208			0.47	0.45	0.39	0.50
0.40	0.44642			0.50	0.50	0.51	0.60
0.50	0.47510			0.61	0.54	0.65	0.71
0.60	0.50378			0.65	0.65	0.65	0.70
0.65	0.51812			0.60	0.55	0.65	0.75

5.7 Triple Drought–Solar Irradiation Model (TDSIM)

Physically some relations can be found among drought magnitude, solar radiation, and sunshine duration. In the dry (wet) season solar radiation and sunshine duration increase (decrease) to their highest (lowest) degree. For cloudy skies, sunshine duration is lowest and rainfall amount increases to a higher level. In conditions of maximum *clearness*, sunshine duration is the highest and generally there is no significant rainfall. If dry periods continue for several successive seasons then *drought* occurs.

The main purpose of the *triple drought–solar irradiation model* (TDSIM) is to combine again three related variables and examine their common behavior by contour maps. In addition, linear model relations are also derived between two variables. The least square and Kriging methods are used for the preparation of the TDSIM. The least square method is used to find the AM parameters and then the third variable *Z-score* concept is taken into consideration for map preparation by the Kriging approach (Journel and Huijbregts 1989; Matheron 1965). This new approach gives not only some climatic variations but also helps to estimate drought intensity depending on sunshine duration and solar radiation intervals.

The Z-score is suggested and designed to quantify the precipitation deficit for multiple time scales, which reflect the impact of drought on availability of different water resources. *Z-scores* are sometimes called "*standard scores*." In every normal distribution, the distance between the mean and a given Z-score cuts off a fixed proportion of the total area under the curve. Statisticians have provided tables indicating the value of these proportions for each possible Z-score (Benjamin and

5.7 Triple Drought–Solar Irradiation Model (TDSIM)

Table 5.5 Standardized precipitation index categories

SPI values	Category
0 to −0.99	MID
−1.00 to −1.49	MOD
−1.50 to −1.99	SED
< −2.00	EXD

Cornell 1970). The Z-score is simply a standardization of a given time series, X_i as $X_1, X_2, \ldots, X_n$, where the standardized series, x_i is calculated as

$$x_i = \frac{X_i - \overline{X}}{S_x} \quad (i = 1, 2, \ldots, n) \tag{5.18}$$

where $\overline{X}$ is the arithmetic mean, and S_x is the standard deviation. A *deficit* occurs at any time when the Z-score is continuously negative. The accumulated magnitude of deficits is referred to as the drought magnitude, and it is the positive sum of the Z-score for all the months within a drought event.

In papers by McKee et al.(1993, 1995) only empirical calculations of drought descriptions such as *mild* (MID), *moderate* (MOD), *severe* (SED), and *extreme* (EXD) drought cases are calculated and accordingly the classifications are done crisply at a single site according to Z-score categories as in Table 5.5.

Similar to standardization by Eq. 5.17 solar radiation and sunshine duration ratios can be standardized as follows:

$$h_i = \frac{(H/H_0)_i - \overline{(H/H_0)}}{S_{H/H_0}} \quad (i = 1, 2, \ldots, n) \tag{5.19}$$

and

$$s_i = \frac{(S/S_0)_i - \overline{(S/S_0)}}{S_{S/S_0}}, \tag{5.20}$$

respectively. Here n is the number of data, and h_i and s_i represent standardized solar radiation and sunshine duration values, respectively. There are physical relations between sunshine duration, solar radiation, and the Z-score indicating drought and wet conditions (see Fig. 5.15).

The following rules can be deduced logically from the simultaneous consideration of these three variables:

1. If sunshine duration is lower than the mean value, ($s_i < 0$) and the Z-score > 0, then wet climatic conditions occur.

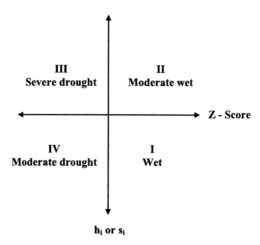

Fig. 5.15 General drought–solar radiation classification

2. If the s_i and Z-score values are greater than zero, then moderate wet climatic conditions occur.
3. If $s_i > 0$ and the Z-score < 0, then severe drought conditions occur depending on the magnitude.
4. Lastly, if $s_i < 0$ and the Z-score < 0 then moderate cloud conditions occur.

The AM and TDSIM methodology are presented for two different climate zones, at İstanbul and Ankara (see Fig. 4.6 and Table 4.3). These two stations have the AM expressions as follows:

$$\frac{H}{H_0} = 0.36\frac{S}{S_0} + 0.28 R^2 = 0.81, r = 0.90 \quad (5.21)$$

and

$$\frac{H}{H_0} = 0.33\frac{S}{S_0} + 0.40 R^2 = 0.83, r = 0.92, \quad (5.22)$$

respectively. The TDSIMs for each station are shown in Fig. 5.16. The İstanbul station lies in the northwestern part of Turkey which is characterized by a modified Mediterranean type of climate with influences from the Black Sea maritime and Balkan continental effects. Consequently, winters are cold and summer seasons are rather warm with long hours of sunshine and high RH. The triple solar irradiation relationship for this city is shown in Fig. 5.16a.

Ankara is in the central Anatolian peninsula and has completely continental climatic effects with dry air movements because humidity-laden air masses from the Black Sea in the north and the Mediterranean Sea in the south leave their moisture along the hills that look toward these water bodies, and the air around Ankara is rather dry with low humidity values.

5.7 Triple Drought–Solar Irradiation Model (TDSIM)

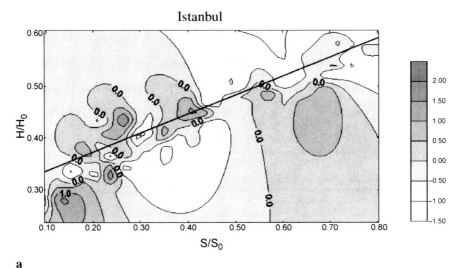

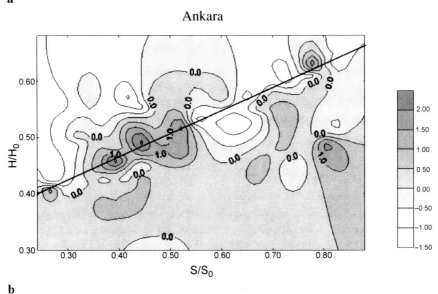

Fig. 5.16 a,b. TDSIMs for **a** İstanbul and **b** Ankara

The TDSIM gives not only climatic variability but also estimates of some solar engineering variables in the AM. If S/S_0 (H/H_0) is known using a and b parameters in Eq. 5.21 and 5.22 then H/H_0 (S/S_0) can be estimated easily. By using solar radiation and sunshine duration ratios, drought magnitude can be estimated with TDSIM. Solar radiation values evaluated by the AM are given in Table 5.6.

This approach can be used also for drought intensity estimation by considerations from Table 5.5. For different S/S_0 and H/H_0 values different drought intensities

Table 5.6 Solar radiation estimation by using the Angström equation

Observed S/S_0	H_M/H_A	Angström estimation H_M/H_A	
		Adana	Ankara
0.300	0.300	0.418	0.405
0.400	0.400	0.446	0.439
0.500	0.500	0.475	0.472
0.600	0.600	0.504	0.506
0.700	0.700	0.532	0.540
0.800	0.800	0.561	0.573

Table 5.7 Drought magnitude estimation by using the Angström equation

Observed S/S_0	H_M/H_A	Drought category	
		Adana	Ankara
0.3	0.3	0.5	1
0.4	0.4	0	1.5
0.5	0.5	0	0
0.6	0.6	0	−1
0.7	0.7	2	−1
0.8	0.8	–	−1

can be estimated, which give different drought magnitudes for different stations (Table 5.7). Hence, drought intensity can be estimated by using sunshine duration and solar radiation ratio ranges.

5.8 Fuzzy Logic Model (FLM)

There are ambiguities and vagueness in solar radiation and sunshine duration records during a day. A *fuzzy logic* (FL) algorithm can be devised for tackling these uncertainties and estimating the amount of solar radiation. The main advantage of *fuzzy logic models* (FLM) is their ability to describe the knowledge in a descriptive human-like manner in the form of simple logical rules using linguistic variables only. The AM or any other type of regression equations are replaced by a set of fuzzy rule bases. There are several implied assumptions in all the model formulations as follows:

1. In many applications without considering the scatter diagram of H versus S, automatically a linear regression line is fitted to data at hand according to a linear model where coefficients depend vaguely on the variations in the sunshine duration. It is the main purpose of this section to provide a simple technique whereby uncertainties in the process of solar radiation and sunshine duration measurements are processed linguistically by means of *fuzzy sets*.
2. A linear model provides estimations of the global radiation on a horizontal surfaces at the level of the earth, but unfortunately it does not give clues about the

normal incidence and tilted surface global radiation because diffuse and direct radiation components do not appear in this linear model.
3. Most of the formulations relate the global irradiation to the sunshine duration by ignoring some of the meteorological factors such as the RH, maximum temperature, air quality, latitude, elevation above mean sea level, *etc*. Each one of these factors contributes to the relationship between H and S and their neglect introduces some errors in the estimations. For instance, the AM assumes that if all of the other meteorological factors are constant then the global horizontal irradiation is proportional to the sunshine duration only. The effects of other meteorological variables appear as deviations from the straight line fit on a scatter diagram. In the FL approach, there are no model parameters but all the uncertainties and model complications are included in the descriptive fuzzy inference procedure in the form of IF-THEN statements.
4. The physical meanings of the model coefficients are not explained in most of the application studies but only the statistical regression fit and parameter estimations are obtained and then incorporated into the relevant formulations for the global irradiation estimation from the sunshine duration records.
5. Dynamic responsive behavior of the system is not considered at all due to the complexities since any regression technique is based on restrictive assumptions (Chap. 4, Sect. 4.4.1). However, in the FLM there are no assumptions involved in the global irradiation estimation from the sunshine duration data. This is because the regression method does not provide dynamic estimation of the coefficients from available data. Furthermore, a critical look at recent interpretations of the AM approach is presented by Gueymard *et al.* (1995) concerning global solar radiation prediction. They have put forward some thoughts and new elements in order to improve the coefficient estimates in the AM. Among their question is "Can statistical and stochastic models be developed and used to supplement linear radiation models?" Herein, the answer to this question is presented by the introduction of the FLM.

5.8.1 Fuzzy Sets and Logic

The concept of *"fuzzy sets"* was introduced by Zadeh (1965) who pioneered the development of fuzzy logic instead of the Aristotelian logic of two possibilities only. Unfortunately, this concept was not welcome into the literature since many uncertainty techniques such as probability theory, statistics, and stochastic processes were commonly employed at that time. FL has been developed since then and it is now used for automatic control of commercial products such as washing machines, cameras, and robotics. Many textbooks provide basic information on the concepts and operational fuzzy algorithms (Tagaki and Sugeno 1985; Tanaka and Sugeno 1992; Wang 1997; Zadeh 1968, 1971; Zimmerman 1991). The key idea in FL is the allowance of partial belonging of any object to different subsets of the universal set instead of belonging to a single set completely. Partial belonging to a set can be described numerically by a membership function (MF) which assumes values between 0 and 1.0 inclusive. For instance, Fig. 5.17 shows typical MFs for small, medium, and large class sizes in a universe, U.

Hence, these verbal assignments are the fuzzy subsets of the universal set. In this figure set values with less than 2 are definitely "small," those between 4 and 6 are certainly "medium," and values larger than 8 are definitely large. However, intermediate values such as 2.2 are in between, that is, it partially belongs to subsets "small" and "medium." In fuzzy terminology 2.2 has an MF value of 0.9 in "small" and 0.1 in "medium" but 0.0 in "large."

The literature is rich with reference concerning the ways to assign MFs to fuzzy variables. Among these ways are intuition, inference, rank ordering, angular fuzzy sets, neural networks, genetic algorithms, inductive reasoning, *etc.* (Şen 2000). In particular, the intuition approach is used rather commonly because it is simply derived from the capacity of humans to develop MFs through their own innate intelligence and understanding. Intuition involves contextual and semantic knowledge about an issue; it can also involve linguistic truth values about this knowledge (Ross 1995).

Fuzzy MFs may take many forms but in practical applications simple straight line functions are preferable. Especially, triangular functions with equal base widths are the simplest possible ones. For instance, in Fig. 5.18 the whole universe of temperature, T, space is subdivided into four subsets with verbal attachments "cold," "cool,"

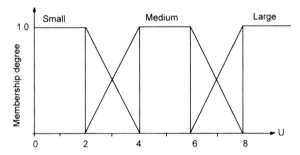

Fig. 5.17 Fuzzy subsets

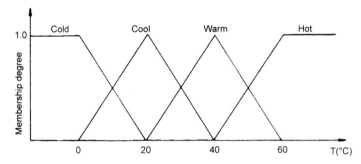

Fig. 5.18 MFs for the fuzzy linguistic word "temperature"

"warm," and "hot," associated with different MFs. Of course, these membership functions are a function of context and the researcher developing them.

The solar radiation at the earth's surface is a random process and therefore it involves uncertainty. Furthermore, if the form of uncertainty happens to arise because of imprecision, ambiguity, or vagueness then the variable is fuzzy and can be represented by an MF.

5.8.2 Fuzzy Algorithm Application for Solar Radiation

A detailed account of the fuzzy sets and logic is given by Şen (1998) for the solar radiation estimation. He used fuzzy logical propositions in the forms of IF-THEN statements. Among a multitude of propositions, two of them are given here below for the sake of argument:

IF sunshine duration is "long" THEN the solar radiation amount is "high."
IF sunshine duration is "short" THEN the solar radiation amount is "small."

In these two propositions solar energy variables of sunshine duration and solar radiation are described in terms of linguistic variables such as "long," "high," "short," and "small." Indeed, these two propositions are satisfied logically by a simple AM. These linguistic variables are only a certain part set of the whole variability domain, *i.e.*, of the full set. It can be understood from this argument that a set of relationships is sought between two variables as exemplified in Fig. 5.19.

This figure shows the architecture of a two-variable fuzzy proposition collection. For our purpose, the first three boxes on the same line represent sunshine duration linguistic words, with the second line three words for the solar radiation. Hence, it is possible to infer $3 \times 3 = 9$ different IF-THEN propositions from Fig. 5.19. The question still remains in this figure is whether the boundaries between the linguistic words in each line are distinct from each other or there may be some overlaps. Logically, it is not possible to draw crisp boundaries between subsequent words. For this purpose, Fig. 5.19 can be rendered into a more realistically valid architectural form

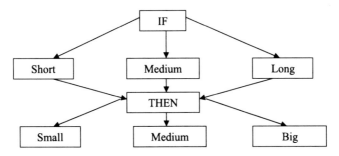

Fig. 5.19 Crisp boundary linguistic words and relationships

as shown in Fig. 5.20 where there are interferences (shaded areas) between the sunshine duration (solar radiation) linguistic words on the same line. The overlapping areas between the atomic words indicate fuzzy regions.

It is also logical to think that as the linguistic word domain moves away from the interference locations they represent more of the linguistic word meaning. For instance, medium sunshine duration linguistic word has two interference regions and, therefore, one can assume comfortably that the middle locations in the "medium" word domain are more genuine mediums. This last statement reflects a triangular type of medium belongingness to the medium region. On the contrary, the words that are located on both sides of the line, such as short and long or small and big, have only one interference region. This means that the belongingness into these words will increase as one moves away from the interference region. Likewise, this gives again the impression of a triangular belongingness but with its greatest belonging at the far edges from the interference regions. Such belongingness is attached with certain numbers that vary between zero and one (Zadeh 1965). In such a terminology zero represents non-belongingness to the word concerned, whereas one corresponds to the full belongingness. These belongingness numbers are referred to as the membership degree (MD) in the fuzzy sets theory. After all these discussions, it becomes evident that the new architecture of the logical propositions will appear as in Fig. 5.21.

Solutions with the architectural form in Fig. 5.20 are already presented by Şen (1998) for solar radiation estimation. In such an approach there is no mathematical equation included. However, in engineering applications simple and linear equations are sought for rapid calculations. For this purpose, the architecture in Fig. 5.21 can be changed into the one in Fig. 5.22 with crisp mathematical forms after the THEN part of the logical propositions (Takagi and Sugeno 1985). In FL terminology, the premises of the productions are vague in terms of fuzzy subsets whereas the consequent parts are adopted in the form of the simplest linear partial mathematical equations.

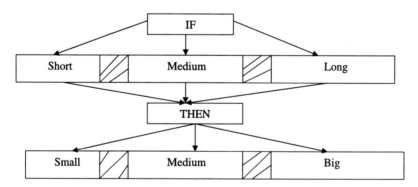

Fig. 5.20 Fuzzy boundary linguistic words and relationships

5.8 Fuzzy Logic Model (FLM)

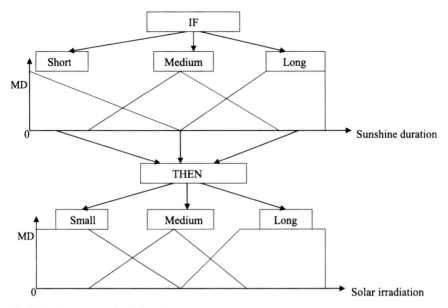

Fig. 5.21 Fuzzy sets and relationships

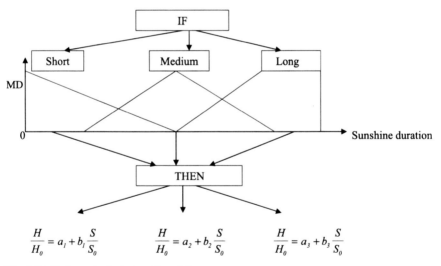

Fig. 5.22 Mathematical relationships of the consequent parts

The mathematical formulations in the consequent part are adopted similar to the AM (Chap. 4). Although the AM is globally fitted to the S/S_0 versus H/H_0 scatter diagram, the architecture in Fig. 5.22 provides a piece-wise linear approach. These two cases are shown representatively in Fig. 5.23.

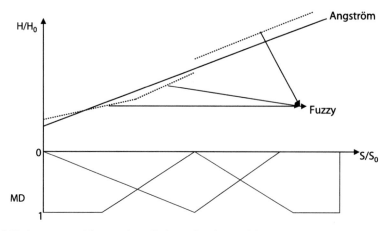

Fig. 5.23 Angström and fuzzy solar radiation estimation models

It is possible to write down the three possible propositions that emerge from Fig. 5.22 as

IF sunshine duration is "short" THEN $\dfrac{H}{H_0} = a_1 + b_1 \dfrac{S}{S_0}$

IF sunshine duration is "medium" THEN $\dfrac{H}{H_0} = a_2 + b_2 \dfrac{S}{S_0}$

IF sunshine duration is "long" THEN $\dfrac{H}{H_0} = a_3 + b_3 \dfrac{S}{S_0}$

Hence, the relationship between the radiation and sunshine duration is deduced from the measurements linguistically. The application of the FLM is performed for two different locations, İstanbul and Ankara in Turkey as shown in Fig. 4.6. Without dividing by H_0 and S_0 salient features of H and S scatter diagram are plotted in Fig. 5.24 for the purpose of fuzzy irradiation estimation. To this end, the extraterrestrial solar radiation, H, and sunshine duration hours are first fuzzified into fuzzy subsets so as to cover the whole range of changes.

The sunshine duration is considered at the maximum for 12 h and its subdivision into 7 subsets as S_1, S_2, S_3, S_4, S_5, S_6, and S_7 is considered to have triangular MFs represented in Fig. 5.25.

However, subsets of fuzzy changes in the solar radiation domain will depend on the location and elevation of the station and accordingly fuzzy partitions will be different for different sites. Since solar irradiance data are continuous, and change slightly and smoothly over very large distances, the fuzzy partitions of solar irradiance are expected to be slightly different for different sites. The reason for adopting seven, contrary to what is recommended as a rule of thumb which in practice is five, is due to the error minimization in the estimation. Once the *fuzzy rule base* inference machine is set up it is straightforward to play with the number of fuzzy partitions on the computer until the best fit is obtained. Of course, the domain of radiation

5.8 Fuzzy Logic Model (FLM)

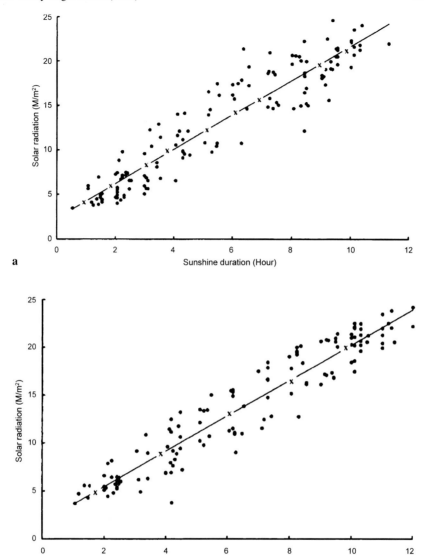

Fig. 5.24 a,b. H versus S scatter diagrams at stations with crosses indicating fuzzy solutions. **a** İstanbul. **b** Ankara

change is observed from the past records and at each station the irradiation values assume 600 cal/cm^2 per month. Due to different latitudes there are three slightly different fuzzy set partitions for İstanbul and Ankara as shown in Fig. 5.26a and b, respectively. The solar radiation fuzzy sets are labeled as H_1, H_2, H_3, H_4, H_5, H_6, and H_7 in increasing magnitude.

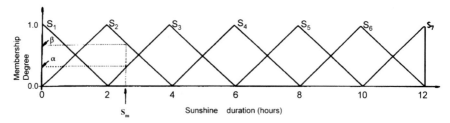

Fig. 5.25 Representative fuzzy subsets for sunshine duration

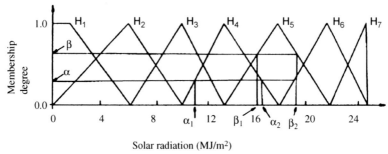

a

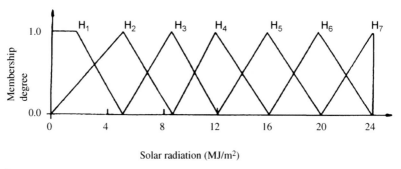

b

Fig. 5.26 a,b. Fuzzy subsets for radiation at different stations. **a** İstanbul. **b** Ankara

The irradiation and sunshine duration fuzzy subsets are combined with each other through the following fuzzy rule bases:

IF S is S_i and S_{i+1} THEN H is H_i and H_{i+1} $(i = 1, 2, 3, 4, 5, 6)$.   (5.23)

Such a fuzzy rule base is not used in any previous study in the literature since the consequent part is in terms of two successive fuzzy subsets from the irradiation domain. For a given sunshine duration measurement, S_m, there are two successive sunshine duration fuzzy subsets (see Fig. 5.25). Once the fuzzy subsets of irradiation

5.8 Fuzzy Logic Model (FLM)

and sunshine duration data are recorded, it is possible to perform the application for the estimation of irradiation amount from a given sunshine duration measurement through the following steps:

1. Locate the measured sunshine duration amount S_m on the horizontal axis in Fig. 5.25. It is possible to find two successive fuzzy subsets, for instance, S_2 and S_3.
2. Find the MDs, namely, α and β corresponding to these two successive sunshine duration fuzzy subsets. It is important to notice at this stage that by definition $\alpha + \beta = 1.0$.
3. Enter the radiation fuzzy subset domain by considering α and β membership degrees as shown, for instance, in Fig. 5.26a.
4. Since α in Fig. 5.25 came from S_4, in Fig. 5.26a it should yield two values, namely α_1 and α_2 from the corresponding H_4 fuzzy subset in the irradiation domain. Likewise, β_1 and β_2 are obtained from H_5 as shown in the same figure. In fact, H_4 and H_5 together present the fuzzy consequent, *i.e.*, answer to the irradiation estimation in the form of a fuzzy subset union as in Fig. 5.27.
5. For the defuzzification of the fuzzy set in Fig. 5.27, first of all the arithmetical averages of the lower and upper values from each fuzzy subsets are calculated as

$$\overline{\alpha} = \frac{\alpha_1 + \alpha_2}{2} \text{ and } \overline{\beta} = \frac{\beta_1 + \beta_2}{2}. \quad (5.24)$$

6. The radiation estimation value, H_e, is calculated as the weighted average of these two simple arithmetic averages as

$$H_e = \alpha\overline{\alpha} + \beta\overline{\beta}. \quad (5.25)$$

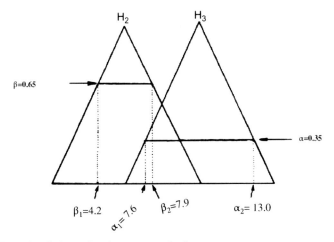

Fig. 5.27 Fuzzy irradiation estimation compound subset

It is to be noticed at this stage that the right hand side of this expression is a function of radiation (see Fig. 5.26a). Hence, it is possible to execute these steps for each sunshine duration measurement which leads to either fuzzy subset estimation in a vague form similar to Fig. 5.27 or after its *defuzzification* by Eq. 5.25 to a single irradiation estimation value. The final results in the form of defuzzified irradiation estimations are shown in Fig. 5.24 by crosses. It is obvious that these crosses lie within almost the central parts of scatter diagram for any given sunshine duration measurement. Hence, the proposed method of fuzzy estimation leads to irradiation estimations either in a vague form similar to fuzzy subset in Fig. 5.27 or as a defuzzified value.

5.9 Geno-Fuzzy Model (GFM)

The FLM accounts for the possible local non-linearity in the form of piece-wise linearization. The parameters estimation of this model can be achieved through the application of the *genetic algorithm* (GA) technique. The fuzzy part of the model provides treatment of vague information about the sunshine duration data whereas the genetic part furnishes an objective and optimum estimation procedure.

Evolution by natural selection is one of the most compelling themes of modern science and it has provoked a revolutionary way of thinking about biological systems. This is a form of evolution referred to as the GA that takes place in a computer. In the GAs, selection operates on strings of binary digits stored in the computer's memory and, over time, the functionality of these strings evolves in much the same way that natural populations of individuals evolve. GAs allow engineers to use a computer to evolve solutions over time instead of designing them by hand. An algorithm is the general description of a procedure and a program is its realization as a sequence of instructions to a computer. Although GAs are known primarily as a problem-solving method, they can also be used to study evolution and to model dynamic systems (Goldberg 1989; Haupt and Haupt 1998; Holland 1992).

The basic idea of a GA is very simple. First, a population of individuals is created in a computer as binary strings and then it is evolved with the use of some principles of variation, selection, and inheritance. In its simplest form, each individual in the population consists of a string of binary digits, which is also referred to as the bits, and by analogy to biological systems the string of bits is referred to as the "genotype." Each individual consists only of its genetic material which is organized into one chromosome. Each bit position (set to 1 or 0) represents one gene. The term "bit string" refers to both genotype and the individuals that they define. There are a variety of techniques for mapping bit strings to different problem domains.

The initial population of individuals is usually generated randomly, although this is not necessary. Each individual is tested empirically in an "environment" and is assigned a numerical evaluation of its merit by a fitness function, F, which returns a single number. This constraint is sometimes relaxed so that F returns a vector of

5.9 Geno-Fuzzy Model (GFM)

numbers, and it determines how each gene (bit) of an individual will be interpreted and thus what specific problem the population will evolve to solve.

Once all the individuals in the population have been evaluated, their Fs are used as the basis for selection, which is implemented by eliminating low-fitness individuals from the population and inheritance is implemented by making multiple copies of high-fitness individuals. GA operations such as mutation (flipping individual bits) and cross-over (exchanging substrings of two individuals to obtain two offsprings) are applied probabilistically to the selected individuals in order to produce a new population (or generation) of individuals. The term cross-over is used here to refer to the exchange of homologous strings between individuals although the biological term "cross-over" generally implies exchange within an individual. New generations can be produced either synchronously, so that the old generation is completely replaced, or asynchronously, so that generations overlap.

By transforming the previous set of good individuals, the operators generate a new set of individuals that have a better than average chance of also being good. When this cycle of evaluation, selection and genetic operations is iterated for many generations, the overall fitness of the population generally improves, and the individuals in the population represent improved "solutions" to whatever problem was posed in F.

There are many details left unspecified by this description. For example, selection can be performed in any of several ways. It could arbitrarily eliminate the least fit 50% of the population and make one copy of all the remaining individuals, it could replicate individuals in direct proportion to their fitness, or it could scale the fitness in any of several ways and replicate individuals in direct proportion to their scaled values. Likewise, the cross-over operator can pass on both offspring to the new generation or it can arbitrarily choose one to be passed on, the number of cross-overs can be restricted to one per pair, two per pair, or N per pair. These and other variations of the basic algorithm have been discussed extensively by various authors (Grefenstette 1985, 1987; Goldberg 1989; Schaffer 1984).

Genetic algorithms are powerful search and optimization algorithms based on semblance of natural genetics which are characterized by the following features. The GA is a fixed procedure that would generate reproducible results and it does not need fine tuning necessary by a skilled experimenter along the following steps:

1. A scheme for encoding solutions referred to as chromosomes to the problem
2. A fitness function that rates each chromosome relative to the others in the current set of chromosomes which are referred to as the population
3. An initiation procedure for the population of chromosomes
4. A set of operators which are used to manipulate the genetic composition of the population
5. A set of parameters that provide the initial settings for the algorithm and operators as well as the algorithm's termination condition

The application of the GFM is presented for İstanbul and Ankara stations in Turkey (see Fig. 4.6, Table 4.3). For the application purpose the sunshine duration fuzzy subsets are already presented in Fig. 5.25.

Table 5.8 GA parameters

Sunshine duration fuzzy subset	Station name İstanbul		Ankara	
	a_g	b_g	a_g	b_g
S_1	0.22	0.84	–	–
S_2	0.27	0.43	0.27	0.48
S_3	0.26	0.44	0.33	0.27
S_4	0.32	0.32	0.28	0.37
S_5	0.37	0.24	0.28	0.36
S_6	0.28	0.36	0.39	0.22
S_7	–	–	–	–
AM	0.28	0.36	0.30	0.34

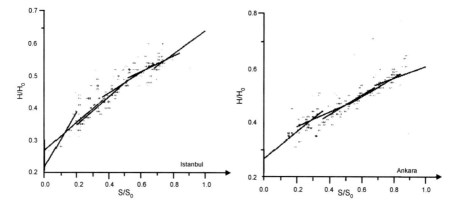

Fig. 5.28 GFMs with straight lines

Table 5.8 shows the GA parameter (a_g and b_g) estimations for each of the seven sunshine duration fuzzy subsets at the two stations.

Figure 5.28 shows each one of the sub-domain straight lines together with the AM straight line.

The AM and its various modifications help to estimate solar radiation from the sunshine duration measurements. Although some of the modifications are in non-linear forms, the estimation of their parameters is achieved through the least squares approach by considering the scatter diagram globally.

5.10 Monthly Principal Component Model (MPCM)

The basis of principal component analysis (PCA) has already been explained in Chap. 4, Sect. 4.7. The scatter diagrams are essential parts prior to the application of PCA and the following questions should be pondered upon in visual and logical interpretation of any solar radiation scatter diagram:

5.10 Monthly Principal Component Model (MPCM)

1. Do the scatter points indicate distinctive mathematical forms? For such an assessment, it is necessary that the scatters should appear around general trends in the forms of linear or non-linear lines.
2. Do the scatter points dispose elongated or circular features? Is there only one feature (cluster) or more in the scatter diagram? (see Fig. 4.18),
3. In the case of a single feature, what is the direction of the most elongated scatter? Similarly, the location of the least elongated direction? In practice, most often these two directions are more or less perpendicular to each other.
4. Are there clustered regions within the scatter diagram? If so, then the regression or any classic model fitting procedure is not suitable and each cluster must be investigated within itself, if possible, independently from other group(s).
5. In the case of small scatters, more distinctive forms appear and it is possible to represent these distinctive features in terms of other convenient Cartesian coordinate systems. A basic and necessary knowledge is that any Cartesian coordinate system has two axes in the plane and they are perpendicular to each other.
6. Is it possible to transform the basic data into some other forms so as to obtain more independent behaviors between the transformed groups? Such independence provides the facility for the researcher to investigate them separately and probably in different contexts.
7. Are the solar radiation (sunshine duration) variation ranges for given sunshine durations (solar radiation) constant? In practice, they are never constant, but theoretical modeling procedures require constant variances along the whole data which cannot correspond appropriately to practical situations.

In the following sequel, the scatter diagram is viewed on a monthly basis with 12 classes. Each class is treated individually by calculating monthly means and variances. This will provide the ability to compare the monthly solar radiation-sunshine duration data variability among the classes.

For the sake of discussion, Fig. 5.29 shows three scatter diagram classes (months) and it is obvious that they have different patterns.

In the same figure, their summary statistics are shown in terms of averages, m_{1H}, m_{2H}, and m_{3H}, on the solar radiation axis and m_{1S}, m_{2S}, and m_{3S} on the sunshine duration axis; σ_{M1}, σ_{M2} and σ_{M3} and σ_{M1}, σ_{m2} and σ_{m3} are the corresponding standard deviations. It is possible to find another set of Cartesian axes such that the maximum (major axis) variance appears along one direction and the minimum (minor axis) along the second axis. Hence, each set of monthly Cartesian axes is represented by an ellipse with the major (minor) axis length equal to twice the maximum (minimum) variance with the center at monthly averages. The comparisons of these ellipses indicate clearly that they are quite distinct from each other. The rotational angle for each month can be calculated according to Sect. 4.7. Finding of these principal axes and the variances along them is referred to in general as the principal component analysis in the literature (Davis 1986), but its two-dimensional version with 12 monthly classifications is referred to as the monthly principal component model (MPCM), which includes the following steps:

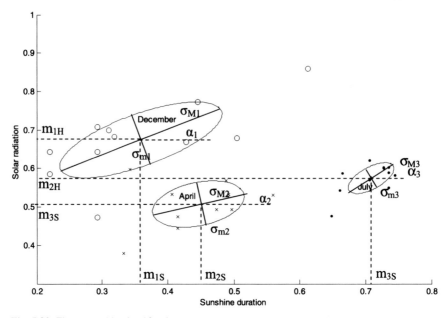

Fig. 5.29 Three monthly classifications

1. Find the arithmetic averages of solar radiation and sunshine duration data for a particular month or season. Let these be indicated for months as m_{iH} and m_{iS} ($i = 1, 2, \ldots, 12$) for solar radiation and sunshine duration data, respectively. Then each pair (m_{iH}, m_{iS}) indicates the center of the corresponding ellipse.
2. Consider any orthogonal axes with their origin at the arithmetic averages.
3. Find the projections of solar radiation and sunshine duration data points on these axes and then calculate the variances along each axis.
4. Repeat the variance calculation for a set of different orthogonal axes at the same central point.
5. Find the maximum variance among all the orthogonal axes and name it as the major axis. Accordingly, the minor axis is perpendicular to this axis.

In this manner, it is possible to identify 12 ellipses as in Fig. 5.30, which correspond to the relevant scatter diagrams.

The MPCM as presented above is applied for five solar radiation sites, which represent different climatologic regions in Turkey (see Fig. 4.6 and Table 4.3). The MPCM parameters for each station and for four months are presented in Table 5.9 Herein, α indicates the angle of the major axes with the horizontal sunshine duration axis; $2\sigma_M$ and $2\sigma_m$ are the lengths of the major and minor axes as indicated in Fig. 5.29.

İstanbul and Ankara monthly ellipse distributions are shown in Fig. 5.31 and 5.32 where three clusters are observable, namely, October–May, June–September, and November, which has the most independent solar radiation and sunshine duration variability.

5.10 Monthly Principal Component Model (MPCM)

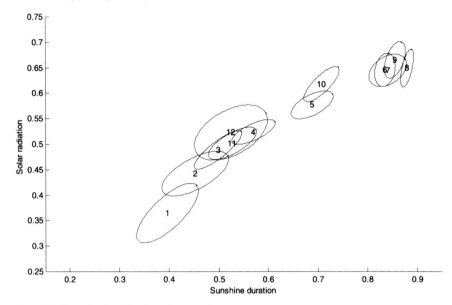

Fig. 5.30 Monthly classification of scatter diagram

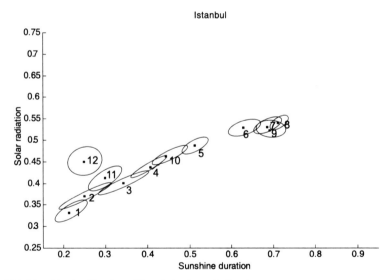

Fig. 5.31 Monthly classification of İstanbul solar radiation data

Ankara station lies almost in the middle of Turkey where a dry continental climate prevails with extremely cold winter periods. If the sequence of numbers $1, 2, \ldots, 12$ corresponding to the monthly sequence of January, February, ..., December, respectively, is followed an increasing limb from 1–8 (January–August) and then a decreasing limb from 9–12 (September–December) are observed. The solar

Table 5.9 MPCM parameters

Month	Parameters	Stations				
		İstanbul	Ankara	Antalya	Trabzon	Kars
January	Average solar irradiation	0.332	0.385	0.544	0.357	0.569
	SD solar irradiation	0.049	0.034	0.108	0.035	0.081
	Average sunshine duration	0.214	0.253	0.588	0.215	0.290
	SD sunshine duration	0.082	0.079	0.151	0.047	0.089
	α	28	20	35	31	40
	σ_M	0.092	0.084	0.183	0.052	0.104
	σ_m	0.026	0.021	0.029	0.025	0.060
April	Average solar radiation	0.434	0.436	0.570	0.384	0.508
	SD solar radiation	0.058	0.045	0.042	0.036	0.060
	Average sunshine duration	0.398	0.475	0.611	0.291	0.446
	SD sunshine duration	0.100	0.094	0.076	0.059	0.069
	α	30	26	25	30	33
	σ_M	0.114	0.104	0.084	0.068	0.075
	σ_m	0.016	0.007	0.025	0.015	0.053
July	Average solar radiation	0.530	0.549	0.634	0.391	0.575
	SD solar radiation	0.038	0.031	0.035	0.036	0.040
	Average sunshine duration	0.682	0.748	0.860	0.337	0.708
	SD sunshine duration	0.064	0.057	0.041	0.076	0.034
	α	27	26	24	25	52
	σ_M	0.070	0.063	0.042	0.084	0.048
	σ_m	0.024	0.015	0.033	0.009	0.023
October	Average solar radiation	0.464	0.526	0.634	0.413	0.566
	SD solar radiation	0.042	0.051	0.025	0.048	0.051
	Average sunshine duration	0.446	0.614	0.762	0.371	0.551
	SD sunshine duration	0.084	0.111	0.055	0.084	0.094
	α	25	24	12	29	21
	σ_M	0.092	0.122	0.056	0.096	0.100
	σ_m	0.019	0.015	0.022	0.014	0.040

radiation and sunshine duration variability is comparatively bigger in November and December. Also, there is an inverse relationship in November.

The monthly solar radiation sunshine distribution in Fig. 5.33 for Antalya is different than Ankara, because it is located in a region in the southern part of Turkey and has a Mediterranean-type climate. Comparatively the monthly scatter diagram has high solar radiation and sunshine duration values.

The sequence of the monthly evolution of these variables does not follow distinctive limbs but rather clusters in Antalya. The first cluster includes the months of November to April whereas the highest cluster has the months of July to September. Between these two clusters May–June is the rising limb transition period and October–November is the recession limb transition period. In December the solar radiation and sunshine duration distribution are independent, because the scatter shape has a circle.

Trabzon station in Fig. 5.34 has an opposite pattern to Antalya station in Fig. 5.33 because it lies along the Black Sea coastal area in the northeastern part of Turkey. There are two distinctive clusters, namely, all months except December.

5.10 Monthly Principal Component Model (MPCM)

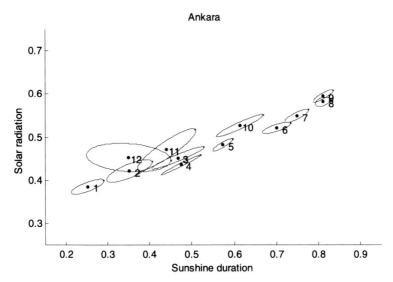

Fig. 5.32 Monthly classification of Ankara solar radiation data

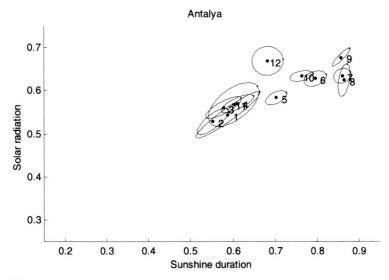

Fig. 5.33 Monthly classification of Antalya solar radiation data

Kars station is located in the northern mountainous region of Turkey and is more than 2000 m above mean sea level. The scatter of monthly partial solar radiation and sunshine scatter has a distinctive feature. It is obvious from Fig. 5.35 that the scatter of the monthly solar radiation and sunshine duration ellipses is comparatively more random than other stations. Such a pattern indicates that the weather conditions at this station are rather unstable and cannot be predicted by any linear or non-linear model.

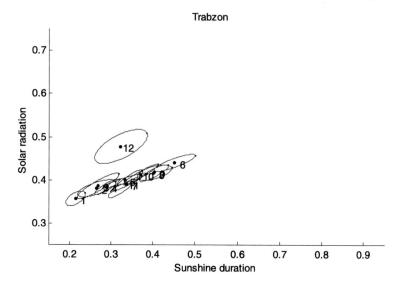

Fig. 5.34 Monthly classification of Trabzon solar radiation data

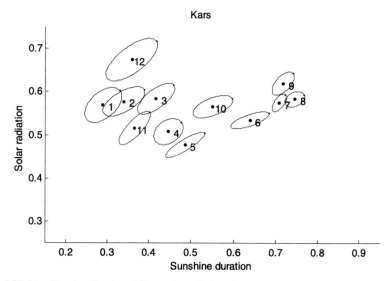

Fig. 5.35 Monthly classification of Kars solar radiation data

The procedure of partial cluster analysis of the solar radiation and sunshine duration predictions are shown as points in Figs. 5.36 and 5.37 for the first and all the years at Kars station, respectively. In the same figures the classic Angström straight line solutions are also presented.

It is clear that the AM solution presents the average behavior of the solar radiation and sunshine duration distribution along a straight line as in Fig. 5.36, which cannot

5.10 Monthly Principal Component Model (MPCM)

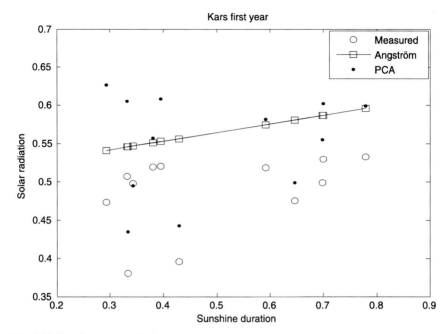

Fig. 5.36 Kars first year predictions

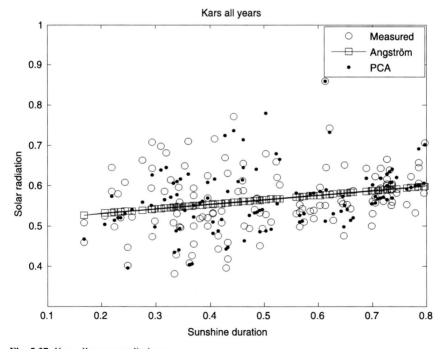

Fig. 5.37 Kars all years predictions

be accepted as a satisfactory solution when visually compared with the measured data points.

It does not represent the scatter around this straight line. On the other hand, for all years in Fig. 5.37, the MPCM predictions follow a far better pattern with the measured values than Angström solution. Hence, not only the average trend of solar radiation and sunshine scatter diagram but additionally the deviations are also simulated, which is very important in solar energy applications.

5.11 Parabolic Monthly Irradiation Model (PMIM)

The main purpose of this section is to present simple models for monthly average hourly global and diffuse radiation estimations. The data are recorded and treated to some extent by Tırıs and Tırıs (1996) for the İstanbul (Gebze, Turkey) solar energy variables measurement station. The measurements are taken on a horizontal surface by using Kipp–Zonen pyranometers through a data logger. The ratio of the monthly total daily diffuse to global radiation amounts are presented in Fig. 5.38.

The ratios vary between 0.33 and 0.55 with a definite periodicity within one year. By definition the diffuse/global ratio changes theoretically between 0 and 1, because the global radiation is always greater than the diffuse radiation amounts. This ratio attains its minimum values in the summer, with the least ratio in July, and the maximum ratios occur in the winter, with the greatest value in December. Physically, the global and diffuse radiation amounts are comparatively bigger in the summer than winter periods. Furthermore, the difference between the global and diffuse radiations in a particular month is greater in summer months than in the winter. The instantaneous global radiations can vary considerably through the day.

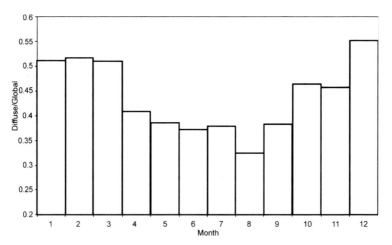

Fig. 5.38 Diffuse to global radiation variations

5.11 Parabolic Monthly Irradiation Model (PMIM)

Tiris and Tiris (1996) have provided scatter diagrams of hourly average global and diffuse radiation changes versus months for different hours within a day. The monthly average hourly global or diffuse radiation changes with hours have random fluctuations around a general trend, which is adopted as the second-order parabola for each month. For the sake of brevity, only one set of these graphs is reproduced in Fig. 5.39.

The mathematical formulation of these changes can be expressed according to a second-order polynomial (PMIM) as

$$I = a_m t^2 + b_m t + c_m \qquad (5.26)$$

where I represents either monthly average hourly global or diffuse radiation amounts, or t indicates time in hours within one day. In this equation a_m, b_m, and c_m are the model parameters that are to be estimated from the available data. By application of the least squares method to the available data the necessary model parameters can be obtained, together with the coefficient of determination, R^2, and the results are presented for the monthly average daily global radiation amounts in Table 5.10. In the meantime, the model curves are shown for different months in Fig. 5.40, together with the scatter of data points.

The R^2 values confirm that the second-order parabola is very suitable as a model of the monthly average hourly global solar radiation change by time.

Similar PMIM parameters for monthly average hourly diffuse radiation are shown in Table 5.11. One can notice from these tables that the values of parameters a_m and c_m are always negative for each month for the global and diffuse radiations amounts, but b_m has positive values. These signs are plausible because, during the daylight hours, substitution of the hour value in any one of the parabolic models does not give a negative radiation amount. More significantly, in order to end the

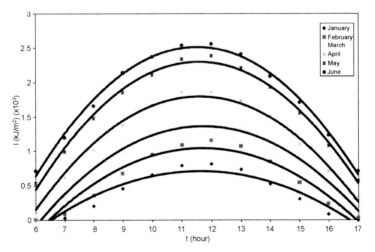

Fig. 5.39 Monthly average hourly solar radiation change with time

Table 5.10 Monthly average daily global radiation parabolic model parameters

Month	Model parameters			Coefficient of determination R^2	$I_m(h)$	I_{max} (kJ/m^2)
	a	b	c			
January	−27.405	636.85	−2990.9	0.8989	11.62	709.0
February	−37.937	887.25	−4145.5	0.9352	11.69	1042.1
March	−44.212	1036.10	−4709.2	0.9782	11.72	1361.0
April	−52.607	1219.90	−5277.0	0.9838	11.59	1795.0
May	−59.818	1384.40	−5711.6	0.9920	11.57	2298.4
June	−61.721	1421.30	−5669.0	0.9932	11.51	2513.4
July	−58.409	1373.00	−5619.5	0.9924	11.75	2449.2
August	−64.616	1500.60	−6352.8	0.9941	11.61	2359.4
September	−62.649	1445.60	−6312.9	0.9864	11.54	2026.3
October	−48.623	1110.70	−4956.5	0.9631	11.42	1386.5
November	−34.355	790.87	−3616.6	0.9183	11.51	935.0
December	−23.448	538.09	−2490.2	0.8837	11.47	596.9
Averages	−47.983	1112.06	−4821.0	0.9599	11.58	1622.7

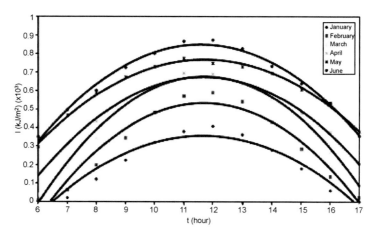

Fig. 5.40 Monthly average hourly diffuse radiation change with time

hour of maximum radiation, t_m, Eq. 5.26 is derived with respect to time and then set equal to zero, which yields

$$t_m = -\frac{b_m}{a_m}. \qquad (5.27)$$

Since the coefficient a_m is always negative in Table 5.10 and 5.11 and b_m is positive, this expression will yield positive times in hours. The amount of maximum radiation, I_{max}, can be formulated by substituting Eq. 5.27 into Eq. 5.26, which leads to

$$I_{max} = -\frac{b_m^2}{4a_m} + c_m. \qquad (5.28)$$

5.11 Parabolic Monthly Irradiation Model (PMIM)

Table 5.11 Monthly average daily diffuse radiation parabolic model parameters

Month	Model parameters			Coefficient of determination R^2	$I_m(h)$	I_{max} (kJ/m^2)
	a	b	c			
January	−13.392	312.20	−1464.2	0.9228	11.66	355.6
February	−19.112	446.86	−2079.4	0.9422	11.69	532.7
March	−21.081	493.27	−2208.9	0.9887	11.70	676.6
April	−16.583	387.14	−1586.3	0.9838	11.67	673.2
May	−13.937	326.52	−1144.3	0.9921	11.71	768.2
June	−16.805	389.24	−1404.0	0.9900	11.58	849.9
July	−20.322	473.14	−1859.7	0.9766	11.64	894.2
August	−17.651	407.98	−1635.3	0.9811	11.56	722.2
September	−20.925	484.92	−2071.0	0.9940	11.59	738.4
October	−21.046	479.58	−2106.9	0.9823	11.39	625.2
November	−15.542	353.42	−1583.0	0.9539	11.37	426.2
December	−12.532	286.94	−1317.7	0.9081	11.45	324.8
Averages	−17.411	403.43	−1705.1	0.9680	11.58	632.2

The changes of a_m, b_m, and c_m parameters for monthly average daily global radiation values are shown in Fig. 5.41.

This figure indicates a systematic variation of parameter a_m within one year, starting from a big value in January, decreasing to a minimum during the summer, and rising again until December. This parameter assumes big values during small radiation months, i.e., during the winter season, and relatively small a_m values appear in big radiation months. For parameter b_m, just the opposite physical interpretation is valid. The behavior of parameter c_m resembles that of parameter a_m. Hence, an increase in the a_m value indicates a decrease in the b_m value but an increase in the c_m value. In order to further confirm these points, Figs. 5.42 and 5.43 indicate the scatter diagram between a_m and b_m as well as a_m and c_m, respectively.

It is obvious that they are related, according to the least squares technique, as

$$a_m = -0.431 b_m - 0.081 \quad (5.29)$$

and

$$a_m = 0.0112 c_m + 6.2467 . \quad (5.30)$$

It is possible to know the values of parameter b_m and c_m provided that a_m is known or *vice versa* as

$$b_m = -23.202 a_m - 1.951 \quad (5.31)$$

and

$$c_M = 89.286 a_M - 557.74 . \quad (5.32)$$

Figure 5.44 represents the monthly average hourly diffuse irradiation model parameter variations, namely, a_m, b_m, and c_m.

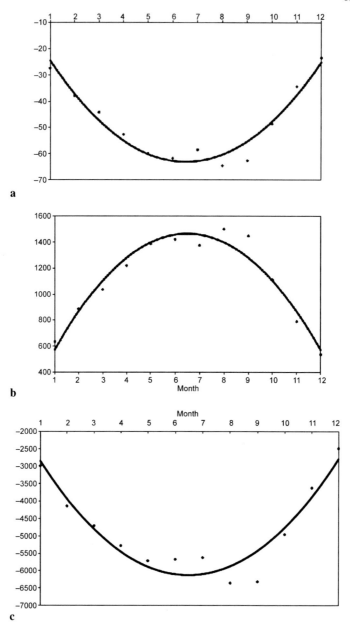

Fig. 5.41 a–c. PMIM parameter variations with time **a** a_m, **b** b_m, and **c** c_m

The comparisons of these figures with their counterparts in Fig. 5.41 indicate that in the case of the monthly average hourly diffuse radiation model, the parameters do not present very definite types of single curvatures, but rather double curvatures

5.11 Parabolic Monthly Irradiation Model (PMIM)

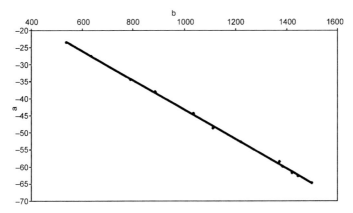

Fig. 5.42 Scatter diagram between a_m and b_m for monthly average hourly global radiation

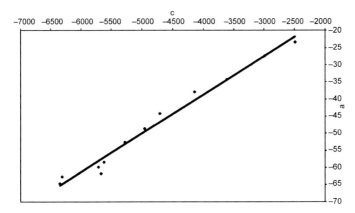

Fig. 5.43 Scatter diagram between a_m and c_m for monthly average hourly global radiation

within one year. Similar relational patterns are observable between the three parameters for monthly average daily diffuse radiation also. Therefore, the scatter diagrams are presented in Figs. 5.45 and 5.46. The resulting equations become

$$a_m = -0.0429 b_m - 0.1169, \tag{5.33}$$
$$a_m = 0.001 c_m - 3.6469, \tag{5.34}$$
$$b_m = -23.31 a_m - 2.725, \tag{5.35}$$

and

$$c_m = 123.457 a_m + 450.234. \tag{5.36}$$

With these sets of equations, if the value of any one of the parameters is estimated then the others can be calculated accordingly.

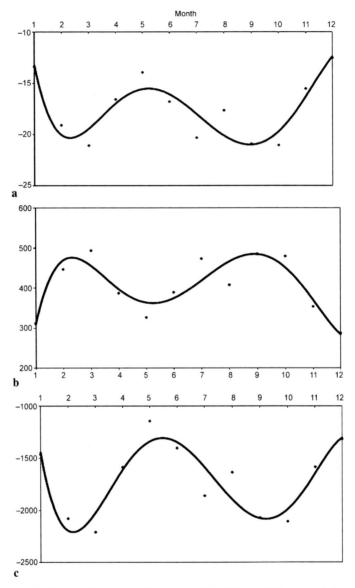

Fig. 5.44 a–c. Monthly variation of monthly average daily diffuse radiation **a** a_m, **b** b_m, and **c** c_m

5.12 Solar Radiation Estimation from Ambient Air Temperature

The designing of solar installations requires the determination of solar radiation incident on the plane of the solar collector. Sloped solar collectors receive direct, diffuse, and reflected solar radiation. In order to calculate the total solar radiation, it is necessary to know their components. The relationship between monthly average

5.12 Solar Radiation Estimation from Ambient Air Temperature

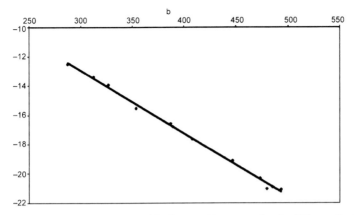

Fig. 5.45 Scatter diagram between a_m and b_m for monthly average hourly diffuse radiation

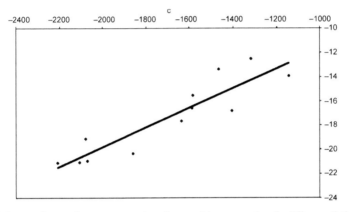

Fig. 5.46 Scatter diagram between a_m and c_m for monthly average hourly diffuse radiation

daily diffuse and global solar radiations incident on horizontal surface, H_d/H is the most significant variable (Chap. 4). This relationship can be found from direct meteorological observations or through an empirical relationship as studied by various researchers (Erbs *et al.*, 1982; Iqbal 1979; Klein 1977). Most of the meteorological stations in many countries record the global and diffuse radiation amounts.

Kenisarin and Tkachenkova (1992) used 34 sets of USSR meteorological station data for establishing such relationships by using 12 monthly daily values of H_d/H and the relationship between the global and extraterrestrial daily solar radiation incident on a horizontal surface, $H/H_0 = K_T$. Linear and non-linear (second- and third-order polynomials) models are taken into consideration for the determination of the correlation type. For instance, the cubic polynomial model with the least squares error is given as follows:

$$\frac{H_d}{H} = 1.191 - 1.783K_T + 0.862K_T^2 - 0.324K_T^3 \quad (0.15 < K_T < 0.80) \ . \quad (5.37)$$

This expression describes evenly the behavior of H_d/H in all investigated latitude intervals. The RMSE of diffuse radiation varies from 11% to 15%.

For the first time Lui and Jordan (1960) developed a graphical relationship between H_d/H and K_T, which is expressed later by Klein (1977) as

$$\frac{H_d}{H} = 1.390 - 4.027 K_T + 5.531 K_T^2 - 3.108 K_T^3 . \tag{5.38}$$

Page (1961) advised the use of the following linear expression for data from ten meteorological stations located between 40°N and 40°S latitudes. This agrees with the linear model in Eq. 4.18 where for the equatorial region $C = 1$:

$$\frac{H_d}{H} = 1.00 - 1.13 K_T . \tag{5.39}$$

On the other hand, Tuller (1976) considered data from four actinometrical stations of Canada and suggested the following expression:

$$\frac{H_d}{H} = 0.84 - 0.62 K_T . \tag{5.40}$$

For latitudes less than 50° the Tuller equation can be recommend whereas for latitudes more than 50° the Iqbal equation is more suitable. Based on data from southern Canadian stations, he recommended the same linear equation with restrictions on K_T as follows:

$$\frac{H_d}{H} = 0.84 - 0.62 K_T \qquad (0.3 < K_T < 0.6) . \tag{5.41}$$

Additionally, with five sets of actinometrical station data from the USA, Collares–Pereira and Rabl (1976) suggested an equation whose coefficient varies with the seasons:

$$\frac{H_d}{H} = 0.775 + 0.347 \frac{\pi}{180} (\theta_{ss} - 90°) - 0.505$$
$$+ 0.261 \frac{\pi}{180} (\theta_{ss} - 90°) \cos\left[2 (K_T - 0.90)\right] \tag{5.42}$$

Based on 12 meteorological stations from India, Modi and Sukhatme (1979) developed a regression equation:

$$\frac{H_d}{H} = 1.4112 - 1.6956 K_T \qquad (0.34 < K_T < 0.73) . \tag{5.43}$$

Models of Lui and Jordan (1960) and Modi and Sukhatme (1979) should be used only for those regions for which they have been proposed.

On the basis of for four US stations, Collares–Pereira and Rabl (1976) and Erbs et al. (1982) recommended the use of the following equations:

$$\frac{H_d}{H} = 1.391 - 3.506 K_T + 4.189 K_T^2 + 2.137 K_T^3 \quad (\theta_{ss} \leq 80° \text{ and } 0.3 \leq K_T \leq 0.8)$$

5.12 Solar Radiation Estimation from Ambient Air Temperature

and

$$\frac{H_d}{H} = 1.311 - 3.022 K_T + 3.427 K_T^2 - 1.821 K_T^3 \quad (\theta_{ss} > 80° \text{ and } 0.3 \leq K_T \leq 0.8). \tag{5.44}$$

Equations 5.37, 5.42, and 5.44 for latitudes less 50° give the non-compensated remainders as -0.025, -0.039, and -0.035, respectively. These remainders are less than for all reference stations which were -0.045, -0.074, and -0.059. The graphical representations of the above correlations are given in Fig. 5.47. It is clear that these correlations approach to different values. Therefore, it is interesting to carry out the statistical comparison of these correlations.

The MBE in Eq. 4.9 and the RMSE in Eq. 4.11 are used for comparison of the present result with other known correlations. The calculation results of the MBE and RMSE from Eq. 5.39 are presented in Table 5.12.

The calculations show that the aforementioned equations give under-estimation of diffuse radiation on all reference stations. In respect to Lui and Jordan (1960), Page (1961), Collares–Pereira and Rabi (1976), and Erbs et al. (1982) the results of the equations agree with the statistical analysis obtained by Ma and Iqbal (1984). More detailed information about the statistical analysis can be found in Kenisarin et al. (1990).

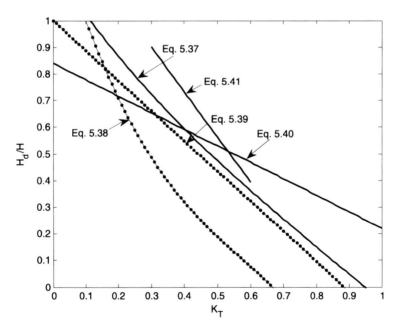

Fig. 5.47 The various types of correlations for H_d/H

Table 5.12 Monthly average daily diffuse radiation estimation on a horizontal surface

Equation	MBE	RMSE
Lui and Jordan (1960)	−0.107	0.132
Page (1961)	−0.043	0.087
Tuller (1976)	0.032	0.092
Iqbal (1979)	−0.017	0.078
Collares–Pereira and Rabl (1979)	0.071	0.149
Modi and Sukhatme (1979)	0.108	0.143
Erbs et al. (1982)	−0.056	0.099
Kenisarin and Tkachenkova (1992)	−0.001	0.075

References

Abdalla YAG, Baghdady MK (1985) Global and diffuse solar radiation in Doha (Qatar). Solar and Wind Technology 2:209

Akınoğlu BG, Ecevit A (1990) Construction of a quadratic model using modified Angström coefficients to estimate global solar radiation. Solar Energy 45:85–92

Barbara S, Coppolino S, Leone C, Sinagra E (1978) Global solar radiation in Italy. Solar Energy 20:431

Benjamin JR, Cornell CA (1970) Probability statistics and decision for civil engineers. McGraw–Hill, New York, pp 684

Benson RB, Paris MV, Sherry JE, Justus CG (1984) Estimation of daily and monthly direct, diffuse and global solar radiation from sunshine duration measurements. Solar Energy 32:523–535

Box GEP, Jenkins GM (1970) Time series analysis forecasting and control. Holden Day, San Francisco

Caughey SJ, Crease BA, Roach WT (1982) A field study of nocturnal stratocumulus II. Turbulence structure and environment. Q J R Meteorol Soc 108:125–144

Collares–Pereira M, Rabl A (1979) The average distribution of solar radiation: correlations between diffuse and hemispherical and between daily and hourly insolation values. Solar Energy 22:155–164

Davies R (1978) The effect of finite geometry on three dimensional transfer of solar irradiance in clouds. J Atmos Sci 35:1712–1725

Davis J (1986) Statistic and data analysis in geology. Wiley, New York

Erbs DG, Klein SA, Duffie JA (1982) Estimation of diffuse radiation fraction for hourly, daily and monthly average global radiation. Solar Energy 28:293–302

Goldberg DE (1989) Genetic algorithms in search optimization and machine learning. Addison–Wesley, Reading

Gopinathan KK (1988) A general formula for computing the coefficients of the correlation connecting global solar radiation to sunshine duration. Solar Energy 41:499–502

Grefenstette JJ (1985) Proceedings of an international conference on genetic algorithms and their applications. NCARAI, Washington, DC and Texas Instruments, Dallas

Grefenstette JJ (1987) Proceedings of the second international conference on genetic algorithms. Erlbaum, Hillsdale

Gueymard CA, Jindra P, Estada–Cajigal V (1995) A critical look at recent interpolations of the Angström approach and its future in global solar irradiation prediction. Solar Energy 54:357–363

Haupt RL, Haupt SE (1998) Practical genetic algorithm. Wiley, New York

Hay JE (1984) An assessment of the mesoscale variability of solar radiation at the Earth's surface. Solar Energy 32:425–434

Holland J (1992) Genetic algorithms. Sci Am, July, pp 44–50

References

Iqbal M (1979) A study of Canadian diffuse and total solar radiation data 1. Monthly average daily horizontal radiation. Solar Energy 22:81–86

Joseph JH, Cahalan RF (1990) Nearest neighbor spacing in four weather cumulus as inferred from Landsat. J Appl Meteor 29:793–805

Journel AG, Huijbregts CJ (1989) In: Mining Geostatistics. Academic, New York, p 600

Kenisarin MM, Tkachenkova NP (1992) Estimation of solar radiation from ambient air temperature. Appl Solar Energy 28:66–70

Kenisarin MM, Tkachenkova NP, Shafeev AI (1990) On relationship between diffuse and global solar radiation. Gelioteclintka 26:72

Klein SA (1977) Calculation of monthly average insolation on tilted surface. Solar Energy 19:325–329

Lui BYH, Jordan RC (1960) The interrelationship and characteristic distribution of direct, diffuse and total solar radiation. Solar Energy 4:1–19

Ma CCV, Iqbal H (1984) Statistical comparison of solar radiation correlations. Monthly average global and diffuse radiation on horizontal surface. Solar Energy 33:143–148

Matheron G (1965) In: Les variables regionalisees et leur estimation. Masson, Paris, p 306

McKee TB, Doesken NJ, Kleist J (1993) The relationship of drought frequency and duration to time scales. Preprints, 8th conference on applied climatology (Anaheim, California), pp 179–184

McKee TB, Doesken NJ, Kleist J (1995) Drought monitoring with multiple time scales. Preprints, 9th conference on applied climatology (Dallas, Texas), pp 233–236

Modi V, Sukhatme SP (1979) Estimation of daily total and diffuse insolation in India from weather data. Solar Energy 22:407–411

Ögelman H, Ecevit A, Taşemiroğlu E (1984) Method for estimating solar radiation from bright sunshine data. Solar Energy 33:619–625

Page JK (1961) The estimation of monthly ea values of daily total short wave radiation on vertical and inclined surfaces from sunshine records for latitudes 40°N–40°S, Proceedings of the UN conference on new sources of energy, paper no. 598, 4, pp 378–390

Rietveld MR (1978) A new method for estimating the regression coefficients in the formula relating solar radiation to sunshine. Agric Meteorol 19:243–252

Ross JT (1995) Fuzzy logic with engineering applications. McGraw–Hill, New York

Şahin A, Şen Z (1998) Statistical analysis of the Angström formula coefficients and application for Turkey. Solar Energy 62:29–38

Schaffer JD (1984) Proceedings of the 3rd international conference on genetic algorithms, Los Altos, CA. Kaufmann, San Francisco

Şen Z (1998) Fuzzy algorithm for estimation of solar irradiation from sunshine duration. Solar Energy 63:39–49

Şen Z (2000) Mühendislikte bulanık mantık ile modelleme ilkeleri (Fuzzy Logic Modeling Principles in Engineering) (in Turkish). Su VakfıYayınları, İstanbul

Şen Z (2001) Angström equation parameter estimation by unrestricted method. Solar Energy 71:95–107

Şen Z (2007) Simple nonlinear solar irradiation estimation model. Renewable Energy 32:342–350

Şen Z, Öztopal A (2003) Terrestrial irradiation: sunshine duration clustering and prediction. Energy Convers Manage 44:2159–2174

Şen Z, Şahin AD (2000) Solar irradiation polygon concept and application in Turkey. Solar Energy 68:57–68

Suleiman SS (1985) Dependence of solar radiation on local geographical factors. Gehotekhnika 21:68

Takagi T, Sugeno M (1985) Fuzzy identification of systems and its applications to modeling and control. IEEE Trans Systems Man Cybern 15:116–132

Tanaka KI, Sugeno M (1992) Stability analysis and design of fuzzy control systems. Fuzzy Sets Systems 45:135–156

Tırıs M, Tırıs C (1996) Correlations of monthly-average daily global, diffuse and beam radiations with hours of bright sunshine in Gebze, Turkey. Energy Convers Manage 37:1417–1421

Tuller SE (1976) The relationship between diffuse, total and extraterrestrial solar radiation. Solar Energy 18:259–263
Wahab AM (1993) New approach to estimate Angström coefficients. Solar Energy 51:241–245
Wang LX (1997) A course in fuzzy systems and control. Prentice–Hall, London
Zabara K (1986) Estimation of the global solar radiation in Greece. Solar Energy Wind Technol 3:267
Zadeh LA (1965) Fuzzy sets. Informat Control 8:338–353
Zadeh LA (1968) Fuzzy algorithms. Informat Control 12:94–102
Zadeh LA (1971) Towards a theory of fuzzy systems. In: Kalman RE, DeClaris N (eds) Aspects of network and system theory. Holt, Rinehart and Winston, New York
Zimmermann HJ (1991) Fuzzy sets theory and its applications. Kluwer Academic, Boston

Chapter 6
Spatial Solar Energy Models

6.1 General

In the previous sections solar radiation modeling is discussed on a given single site. However, in practical solar energy assessment studies, it is necessary to have spatial (multiple sites) solar energy estimation procedures. The spatial solar energy distribution depends not only on the meteorological effects such as the clouds, aerosols, *etc.*, but also on the topography, geographic location, and land use and soil type. In practice, single site measurements at a set of irregular locations are available. Logically, the spatial solar radiation variation at any non-measurement site must be deduced by some scientific methodologies. The spatial weights are deduced through the *regionalized variables* (ReV) theory (Journel and Huijbregts 1989; Matheron 1965), the semivariogram (SV) method (Matheron 1971), and the cumulative SV (CSV) approach (Şen 1989). These methods help to find the change of spatial variability with distance from a set of given solar radiation data and then estimation of the solar radiation value is achieved at any desired site through a weighted average procedure. The number of adjacent sites considered in the weighting scheme is based on the least squares technique which is applied either globally or adaptively. The validity of these methodologies is checked with the cross-validation technique.

The main purpose of this section is to present and develop regional models for any desired point from solar radiation measurement sites. The use of geometric functions, inverse distance (ID), inverse distance square (IDS), SV, and CSV techniques are presented for the solar radiation spatial estimation. Empirical CSVs and standard spatial dependence function (SDF) are developed as an alternative to the classic spatial autocorrelation function of the irradiation data. The SDF is proposed and applied instead of classic weighting functions, which takes into consideration not only the measurement sites' configuration but also the records of solar radiation values at each site.

6.2 Spatial Variability

In general, *spatial variability* is concerned with different values for any property, which is measured at a set of irregularly distributed geographic locations in an area. The aim is to construct a regional model on the basis of measurement locations with records and then to use this model for regional estimations at any desired point within the area.

Solar radiation varies both in time and space, and its sampling is based on the measurement stations' configuration. In many practical applications measured data are seldom available at the point of interest and consequently the only way to transfer the solar radiation data from the measurement sites to the estimation point is through regional interpolation techniques. The spatial variability in irradiation is a function of the season in a year and time interval in the season. Synoptic weather conditions also play a role in the spatial features. The longer the duration the smaller will be the regional variability and as the time interval becomes shorter the variability will increase accordingly. The spatial variability is measured in the most common way through the recorded solar radiation time series at individual points. The relative variability between the stations (the difference of simultaneous values between each pair of stations) is treated commonly by the spatial autocorrelation function, which is used for inter-station dependence based on a set of restrictions (Şen and Habib 2001). Its use is not recommended unless either the irradiation and sunshine duration data abide by the Gaussian (normal) probability distribution function (PDF) or at least they can be transformed into the Gaussian PDF. Hay (1983, 1984) has presented autocorrelation function studies for the North American countries, where the regional variability has not been well documented quantitatively in the literature. Among the questions asked are the following points:

1. What is the optimum number of solar radiation measurement stations in the area considered?
2. How should these stations be arranged so as to describe adequately the radiation climate of the area within which the stations are located?
3. What is the objective procedure that will provide an unbiased estimate of the inter-station dependence between the stations?
4. How should the *radius of influence* be determined, which is defined as the average distance over which the spatial correlation has significant values?

According to Hay (1986) the smaller is the sampling period the smaller will be the correlation coefficient. Solar energy is significant for small durations because solar radiation variation has high frequency components. As the time duration increases the energetic effects of the solar radiation on the climatologic quantities become less because they are averages over longer durations (Chap. 2). Their study of the North American solar radiation records showed a spatial correlation of 0.40 at 250 km for a 1-h period but this increased to 0.98 for a 5-year period. Statistical facts indicate that since over longer time durations summation or averages are calculated, their correlation increases because such averages become abundant in lower frequency components. In meteorological studies, the optimum interpolation

6.2 Spatial Variability

method of *regional dependence* is used to evaluate the error distribution from the solar radiation measurement networks (Gandin 1963). The critical point in such estimation is the use of five measurement sites adjacent to the point of interest. This is a very subjective approach because the fixation of five stations does not have any scientific basis apart from its practical convenience. Besides, such an approach leaves rather large errors in the final estimation especially where spatial variability is large in the mountainous and coastal areas or in areas of widely dispersed measurement stations.

Hay (1979) stated that current analytical programs and observational techniques are not capable of dealing with the substantial variability existing in the solar radiation data. Suckling and Hay (1978) have attempted to show that a synoptic approach of solar radiation regimes may provide a more useful basis for interpretations than calendar periods leading to irradiation climate variability in a region. Accordingly temporal and spatial radiation weather characteristics are associated with the distinctive solar radiation regimes. It is important to develop models that provide a quantitative basis for the assessment of the solar radiation data variability around a given measurement station.

Spatial variability is the main feature of regionalized variables, which are very common in the physical sciences (Cressie 1993). In practical applications, the spatial variation rates of the phenomenon concerned are of great significance in fields such as solar engineering, agriculture, remote sensing, and other earth and planetary sciences. A set of measurement stations during a fixed time interval (hour, day, month, *etc.*) provides records of the regionalized variable at irregular sites, and there are few methodologies to deal with this type of scattered data. There are various difficulties in making spatial estimations originating not only from the regionalized random behavior of the solar radiation, but also from the irregular site configuration. Hence, the basic questions are as follows:

1. How to transfer the influence of the neighboring measurement stations to the estimation point?
2. How to combine these effects to make reliable regional estimates of solar irradiance?

Based on empirical work by Krige (1951) for estimating ore grades in gold mines, the *regionalized variables* (ReV) theory was developed by Matheron (1971). This theory is also termed *geostatistics*, which has been used to quantify the spatial variability. The basic idea in geostatistics is that for many natural phenomena, such as solar radiation, samples taken close to each other have a higher probability of being similar in magnitude than samples taken further apart, which implies a *spatial correlation* structure in the phenomena. Especially, in earth sciences, considerable effort has been directed toward the application of the statistical techniques leading to convenient regional interpolation and extrapolation methodologies (Barnes 1964; Clark 1979; Cressman 1959; Sasaki 1990).

The spatial solar radiation estimation problem has been addressed first by Dooley and Hay (1983) and Hay (1984). They tried to evaluate the errors using solar irradiance data at a number of sites in Canada. The basis of their approach was the *optimal interpolation* techniques as suggested by Gandin (1963) in the meteorology literature. The main interest was to estimate the long-term average of all the sites considered for each month irrespective of any particular year. Systematic interpolation evaluations have been carried out in solar radiation networks by different authors (Hay 1983; Şen and Şahin 2001; Zelenka 1985; Zelenka et al., 1992).

It is possible to prepare solar radiation maps of a region based on a set of measurements at different sites by using basic geostatistical techniques such as semivariograms (SV) and then the Kriging methodology (Journel and Huijbregts 1989). The success of Kriging maps is dependent on the suitability of the theoretical SV to the data at hand. In fact, SVs are the fundamental ingredients in Kriging procedures, because they represent the spatial correlation structure of the phenomenon concerned. There are, however, practical difficulties in the identification of empirical SVs from available data (Şen 1989, 1991).

6.3 Linear Interpolation

The essence of the *spatial interpolation* is to transfer available information in the form of data from a number of adjacent irregular measurement sites to the estimation site through a function that represents the spatial weights according to the distances between the sites (Fig. 6.1).

Generally, changes in the measurement site number or, especially, the location of the estimation site will cause changes in the weightings due to change in the distances. In the *linear interpolation* technique as presented by Gandin (1963) the solar radiation estimation site, S_E, is assumed to be a linear combination of the records at n measurement sites, $S_i (i = 1, 2, \ldots, n)$, which can be expressed as

$$S_E = \frac{\sum_{i=1}^{n} W(r_{i,E}) S_i}{\sum_{i=1}^{n} W(r_{i,E})}, \qquad (6.1)$$

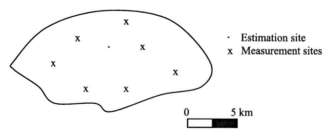

Fig. 6.1 Measurement and estimation sites

6.3 Linear Interpolation

where $r_{i,E}$ and $W(r_{i,E})$ are the distance and the *weighting function* between the i-th solar radiation measurement site and the estimation site. By defining that

$$w_i = \frac{W(r_{i,E})}{\sum_{i=1}^{n} W(r_{i,E})}, \qquad (6.2)$$

it is possible to write Eq. 6.1 as

$$S_E = \sum_{i=1}^{n} w_i S_i, \qquad (6.3)$$

where w_is are the weighting factors which show the contribution from the is-th measurement site. Due to the unbiased estimation requirement, the summation of the weights must be equal to 1 as a restriction:

$$w_1 + w_2 + \ldots + w_n = 1. \qquad (6.4)$$

It is the purpose of any regional method to determine these weights in an optimum manner. Such a regional estimation gives rise to an error, e, which is defined as the difference between the solar irradiance estimation, S_E, and the measured values S_i. The estimation *error variance*, V_e, as defined below, must be minimum:

$$V_e = \frac{1}{n} \sum_{i=1}^{n} (S_E - S_i)^2. \qquad (6.5)$$

The same estimation variance may also be used for *cross-validation*, whereby the measured solar radiation value at site i is considered as if it is not measured (Sect. 6.6.1). The analytical derivation of weightings for the data is found in Cressie (1993). After the substitution of Eq. 6.3 into Eq. 6.5, then taking the partial derivatives with respect to weightings and their equalization to zero provide n equations with n unknown weighting factors and the simultaneous solution yields the weighting factors in terms of the data values. Hence, the *best linear unbiased estimate* (BLUE) of the weightings and consequently the solar radiation estimations are obtained. In this approach the unbiased requirement is not satisfied unless the departures from the ensemble average are used rather than the observations themselves. However, in practical studies the observations are used and therefore the resultant estimates are biased.

Another linear interpolation technique which is very commonly used in many regionalization and spatial estimation problems is the *Kriging* approach which provides BLUE for the spatial interpolations. It is based on the linear estimator as in Eq. 6.3 and minimization of the estimation variance. The Kriging technique as suggested by Matheron (1965) was applied for the first time in earth sciences for ore body recovery in mining, but it has several applications in the atmospheric and hydrological sciences (Delhomme 1978; Journel and Huijbregts 1989). In many practical applications, it is not possible to obtain a representative SV due to either paucity of measurement sites or the regional discontinuity and heterogeneity in the ReV evolution. Bardossy *et al.* (1990a,b) indicated that when sampling sites are inadequate for establishing an SV, Kriging estimates are not reliable.

However, in the classic applications the *inverse distance* (ID) and *inverse distance square* (IDS) methods provide weights, $W(r_{i,E})$, in Eq. 6.1 simply as equal to $1/r_{i,E}$ and $1/r_{i,E}^2$, respectively. Since, weightings do not take into consideration the spatial variation in the solar radiation data the final estimation will be biased. In this sense, they are similar to geometric weighting functions.

6.4 Geometric Weighting Function

The weighting functions are prepared on rational and logical bases without consideration of regional data variability and hence they have the following major drawbacks:

1. They do not take into consideration the natural regional variability features of solar radiation. For instance, in meteorology, the Cressman (1959) weighting function is proposed as

$$W(r_{i,E}) = \begin{cases} \dfrac{R^2 - r_{i,E}^2}{R^2 + r_{i,E}^2} & \text{for } r_{i,E} \leq R \\ 0 & \text{for } r_{i,E} \geq R \end{cases}, \quad (6.6)$$

   where R is the *radius of influence* and it is determined subjectively by personal experience.

2. Although weighting functions are considered universally applicable all over the world, they may show unreliable variability for small areas. For instance, within the same study area, neighboring sites may have quite different weighting functions.

3. Geometric weighting functions cannot reflect the morphology, climatology, and meteorology, which cause the regional variability of the phenomenon. In practical studies, ID, IDS, and geometric weighting functions can be considered only as first approximations in any spatial estimation procedure.

A generalized form of the Cressman model with an extra exponent parameter α is suggested by Thiebaux and Pedder (1987) as

$$W(r_{i,E}) = \begin{cases} \left(\dfrac{R^2 - r_{i,E}^2}{R^2 + r_{i,E}^2}\right)^\alpha & \text{for } r_{i,E} \leq R \\ 0 & \text{for } r_{i,E} \geq R \end{cases}. \quad (6.7)$$

6.4 Geometric Weighting Function

The inclusion of α has alleviated the aforesaid drawbacks to some extent, but its determination still presents difficulties in practical applications. Another form of geometric weighting function was proposed by Sasaki (1960) and Barnes (1964) as

$$W(r_{i,E}) = \exp\left[-4\left(\frac{r_{i,E}}{R}\right)^2\right]. \tag{6.8}$$

Figure 6.2 shows various dimensionless geometric weighting functions used in the atmospheric and earth sciences literature by different researchers, where R indicates the radius of influence and r is the actual distance between any two stations.

All the aforementioned techniques depend on the geometric distances only without consideration of solar radiation records at each site. Hence, none of these techniques are event-dependent but based on logical and geometric concepts. If any one of the relationships in Fig. 6.2 is used, then the estimation will be biased to a certain extent since, for the application of these relationships, it is necessary to determine subjectively the radius of influence. Available weighting functions in the literature are proposed on a logical basis by taking into consideration the site configuration only without experimental verification (Barnes 1964; Cressman 1959; Sasaki 1960; Thiebaux and Redder 1987). In general, the major drawbacks of the weighting functions available in the literature are given by Şen (1997). On the other hand, in the

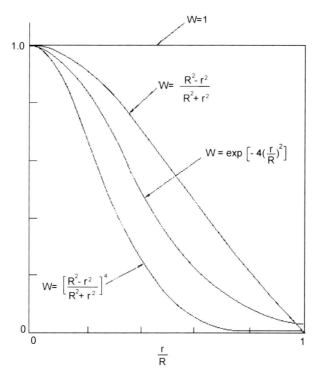

Fig. 6.2 Geometric weighting functions

SV- and CSV-based approaches, similar relationships are determined with the procedures based on the available data set. Hence, they change as a result of changes in the site configuration and ReV data. Experimental SVs and CSVs lead to weighting functions which reflect the regional behavior of the available data.

6.5 Cumulative Semivariogram (CSV) and Weighting Function

The solar radiation measurements are recorded at a set of irregular measurement points within any area at regular time intervals (hourly, daily, and monthly). Solar radiation shows variations in the atmosphere with respect to time and location. The temporal and regional evolutions are controlled by temporal and spatial *correlation structures*. As long as the measurements are at regular time intervals (hourly, daily, and monthly), the whole theory of time series is sufficient in their temporal modeling, simulation, and prediction (Box and Jenkins 1970). The problem is with their spatial measurements at a set of irregular sites, the information transfer from irregular sites to regular grid nodes or to any desired point. Provided that the regional dependence structure is depicted effectively then any future study, such as the numerical estimation at any site based on measurement sites, can be applied successfully.

To this end, *regional covariance* and SV functions are among the early alternatives for the weighting functions that take into account the *spatial correlation* of the phenomenon considered. The former method requires a set of assumptions such as the Gaussian distribution of the regionalized variable. The latter technique, SV, does not always yield a clear pattern of regional correlation structure. Hence, herein the CSV method is used, which does not suffer from these drawbacks. It is proposed by Şen (1989) as an alternative to the classic SV technique of Matheron (1971) with various advantages over any conventional procedure in depicting the *regional variability*, and hence, spatial dependence structure. The CSV is a graph that shows the variation of successive half-squared difference summations with distance. Hence, a non-decreasing CSV function is obtained which exhibits various significant clues about the regional behavior. The CSV provides a measure of regional dependence such that the closer the two sites, the more correlated the regional event and the smaller is the value of the CSV. Prior to the derivation of the experimental CSV weighting functions, some logical points embedded in the aforementioned geometric weighting functions (Fig. 6.2) must be identified:

1. The weighting functions have dimensionless forms by dividing the distances and weights to their respective maximum values (see Fig. 6.2). On both axes the variations are confined between 0 and 1. On the horizontal (vertical) axis is the dimensionless distance (weighting).
2. The maximum (minimum) weight corresponds to minimum (maximum) distances. In the case of wavy field the monotonically decreasing weighting functions with distance are also expected to have waves.

3. The weighting functions decrease monotonically with distance, which is not a necessary requirement in the case of the CSV approach.
4. The steeper is the slope in the weighting function the smaller is the domain of influence along the distance axis.

Solar radiation values are more accurately estimated from local sunshine observations than by assignment from nearby *pyranometric* stations if the latter are more than 20 to 30 km away. Convenient spatial models require distance between any two locations (measurements and estimation sites) for evaluating the regional variability of solar radiation. The CSV can be obtained from a given set of solar radiation data by execution of the following steps:

1. Calculate the distance $d_{i,j}$, $(i \neq j = 1, 2, \ldots, m)$ between every possible pair of measurement sites. For instance, if the number of sample sites is n, then there are $m = n(n-1)/2$ distance pairs.
2. For each distance, $d_{i,j}$, calculate the corresponding half-squared differences, $D_{i,j}$, of the solar radiation data. For instance, if the solar radiation variable has values of S_i and S_j at two distinct sites at distance $d_{i,j}$ apart, then the half-squared difference is

$$D_{i,j} = \frac{1}{2}(S_i - S_j)^2 . \tag{6.9}$$

3. The plot of distances versus half-squared differences yields the experimental SV. However, accumulations of the half-squared differences starting from the smallest distance to the largest leads to the experimental CSV in the form of a non-decreasing function as

$$\gamma(d_{i,j}) = \frac{1}{2}\sum_{i=1}^{n}\sum_{i=1}^{n} D_{i,j} , \tag{6.10}$$

where $\gamma(d_{i,j})$ represents CSV value at distance $d_{i,j}$.
4. Plot $\gamma(d_{i,j})$ values versus the corresponding distance $d_{i,j}$ (Fig. 6.3). The sample CSV functions are free of subjectivity because no *a priori* selection of distance classes is involved in contrast to the analysis as suggested by Perrie and Toulany (1965) in which the distance axis is divided into subjective intervals, and subsequently, averages are taken within each interval which is regarded as the representative value for the interval.

6.5.1 Standard Spatial Dependence Function (SDF)

Prior to the formal derivation of standard *spatial dependence function* (SDF), it is necessary to obtain the experimental CSV, the general features of which are already explained by Şen (1989). By keeping in mind some logical points embedded in the

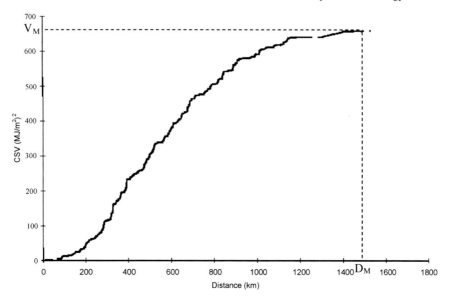

Fig. 6.3 Sample CSVs

geometric weighting functions, the CSV can be converted into an SDF through the following steps:

1. Find two successive CSV values a large distance apart that are almost equal to each other within a certain percentage of relative error, say less than 0.05. This CSV level can be regarded as the maximum value, V_M, (Fig. 6.3).
2. Divide the whole CSV ordinates by V_M and, hence, a dimensionless experimental CSV results as a standard non-decreasing function for the regionalized variable.
3. In order to obtain the experimental SDF, it is necessary to subtract the dimensionless CSV ordinates at every distance from one. The resulting graph is a non-increasing but steadily decreasing function of distance similar to any regional correlation or geometric weighting functions in the literature.
4. The SDF provides a basis for the selection of the weights corresponding to any distance between two sites. Hence, distances on the horizontal axis yield through the SDF, the weights on the vertical axis. The weights are positive and assume values between zero and one. The greater is the distance the smaller the weights and *vice versa*.

For the implementation of the methodology Table 6.1 shows 21 solar radiation measurement stations with measurements in column 2. Figure 6.4 shows the locations of the sites within the area.

It is obvious that there is a rather representative distribution of the sites over the whole study area. The classic experimental SV values from the data yield scatter points as in Fig. 6.5 which are rather haphazard.

6.5 Cumulative Semivariogram (CSV) and Weighting Function

Table 6.1 Measured and estimated solar radiation values and errors

Station	Measurement	Estimation		Relative error (%)	
		Global	IDS	Measurement-Global	Measurement-IDS
(1)	(2)	(3)	(4)	(5)	(6)
1	35.5	32.85	33.44	7.45	5.80
2	29.4	32.78	31.92	11.51	8.57
3	36.8	33.72	30.87	8.36	16.11
4	33.7	34.15	30.77	1.32	8.69
5	35.3	33.33	36.67	5.58	3.88
6	32.5	32.56	35.61	0.19	9.57
7	30.6	33.88	33.24	10.73	8.63
8	30.1	33.52	35.47	11.37	17.84
9	40.1	33.76	32.62	15.81	18.65
10	31.6	33.84	33.24	7.1	5.19
11	34.8	33.32	36.66	4.24	5.34
12	28.6	32.76	30.5	14.55	6.64
13	41.5	33.48	29.85	19.32	28.07
14	33.2	33.51	34.6	0.92	4.22
15	34.3	33.56	37.35	2.31	8.89
16	31	33.82	33.24	9.09	7.23
17	29.6	32.5	34.67	9.78	17.13
18	40.4	33.7	34.04	16.59	15.74
19	28.5	34.08	34.12	19.59	19.72
20	24.4	34.55	32.29	41.62	32.34
21	39.5	32.91	34.06	16.69	13.77
Average	33.4	33.46	33.58	11.15	12.48

The scattering in this figure represents a "variogram cloud" which is the collection of half-squared differences of a ReV for all the possible pairs of points within the data set.

The SV scatter in Fig. 6.5 is added successively starting from the smallest distance leading to the CSV pattern as shown in Fig. 6.6. It can now be used for obtaining a suitable SDF for the data at hand after the execution of steps already mentioned in this section (see Fig. 6.7).

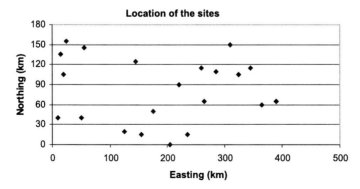

Fig. 6.4 Solar radiation measurement station locations

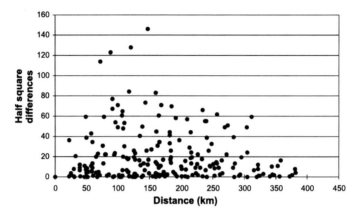

Fig. 6.5 Solar radiation semivariogram

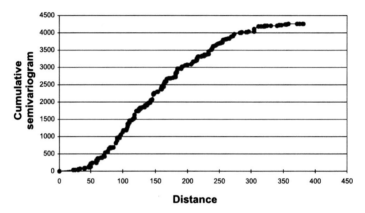

Fig. 6.6 Solar radiation CSV

Similar to all the regional estimation procedures, weighted average formulation as in Eq. 6.3 are used together with the weights obtained from the SDF in Fig. 6.7b.

6.6 Regional Estimation

Regional estimation of solar radiation is essential for economic planning, operation, and maintenance of many solar energy projects. There are different estimation techniques as mentioned before including purely geometric methods (Barnes 1964; Cressman 1959; Thiebaux and Pedder 1987) or distance weightings (Davis 1986) or more efficient geostatistical approaches through semivariogram (SV) techniques such as the Kriging technique (Clark 1979; Journel and Huijbregts 1989; Matheron 1965). Additionally, it is also possible to divide the study area into sub-areas of polygons. Among these are the Thiessen polygons (Thiessen 1912), percentage

6.6 Regional Estimation

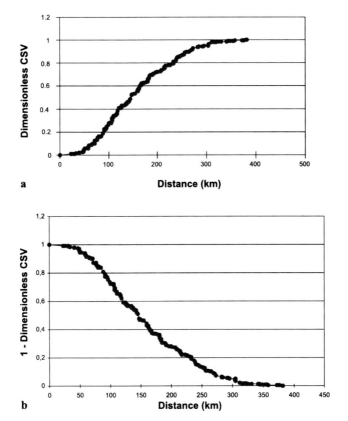

Fig. 6.7 a,b. Solar radiation **a** dimensionless CSV, **b** SDF

weighting polygon (Şen 1998), Delauney triangularization, or other types of regular or irregular polygons.

6.6.1 Cross-Validation

In order to assess the validity of the proposed weighted average procedure, a cross-validation technique is used. According to this, a data value at one site is supposed to be unknown and it is removed from the data set. This removed value is then estimated with the remaining set of data by using the SDF together with Eq. 6.3. This procedure is repeated for all the sites knowing that a datum removed for estimation at its location is put back again in the set for the estimation of another location.

6.6.1.1 Global Estimation

There are two procedures in the estimation of the solar radiation value. In the first one all the sites are considered for their simultaneous contributions and, therefore, in the estimation of any solar radiation value all the distances from this site to others (n-1 sites) are measured from the location map (Fig. 6.4). Subsequently, these distances are entered into Fig. 6.7b on the horizontal axis and the corresponding SDF weights are found from the vertical axis for each distance. In this manner, all the sites are treated equally and hence, instead of measured values, their estimations through the SDF and the cross-validation procedure are calculated. Column 3 in Table 6.1 shows estimated solar radiation values and their corresponding relative errors are calculated as the solar radiation ratio of the absolute difference between the measured and estimated values divided by the measured values multiplied by 100, which are shown in column 5:

$$\text{Relative Error} = 100 \frac{|\text{measured-estimated}|}{\text{measured}}. \quad (6.11)$$

In order to assess visually the correspondence between the measured and estimated values, solar radiations are presented in Fig. 6.8 against the station number sequence along the horizontal axis.

It is obvious that for almost half of the sites, the relative error is more than 10%, which indicates the unsuitability of the global estimation procedure. For extreme solar radiation concentration sites the relative errors are very high. However, the av-

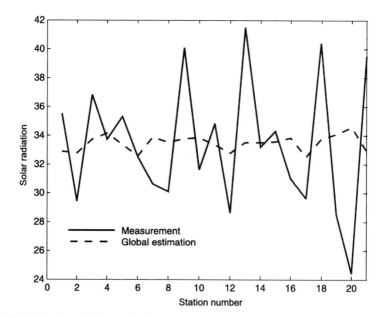

Fig. 6.8 Global solar radiation estimations

6.6 Regional Estimation

erages of measured and estimated values are very close within the 2% relative error band. On this basis, it may be concluded that the proposed procedure yields reasonable values on average but fails to estimate individual site values. This procedure takes into account the contribution of all the sites in the estimation and disregards the concept of the radius of influence.

On the other hand, the results by the IDS are also shown in column 4 and with their relative errors in the last column of Table 6.1. The comparisons of average relative errors in the last row with global approach errors indicate that the latter has superiority because it takes into consideration the actual solar radiation measurement in the estimations.

6.6.1.2 Adaptive Estimation

In order to improve the representativeness of the solar radiation regional estimations at sites, herein, an adaptive technique is suggested which not only estimates the regional value at a site but also provides the number of the nearest sites that should be considered in the best possible regional estimation. Accordingly, the *radius of influence* is defined as the distance between the estimation site and the furthest site within the adjacent sites that are considered in the regional estimation procedure. The following steps are necessary for the application of this adaptive procedure:

1. Take any site for cross-validation and apply Eq. 6.3 by considering the nearest site only. Such a selection is redundant and corresponds to the assumption that, if only the nearest site measurement is considered then the regional estimation will be equal to the same value. This means that in such a calculation the radius of influence is the minimum and equal to the distance between the estimation site and the nearest site.
2. Consider now two of the nearest sites to the estimation site and apply the SDF weighting method according to Eq. 6.3. Consideration of these two sites will increase the radius of influence as the distance between the estimation and the next nearest site and the estimation value will assume the weighted value of the two nearest sites. Since, the weights and measurements are positive numbers, the estimated value will lie between the measurements. There will be a squared estimation error as the square of the difference between measured and estimated values.
3. Repeat the same procedure now with the three nearest stations and calculate the square error likewise. Subsequently, it is possible to continue with $4, 5, \ldots, (n-1)$ nearest sites, and for each case to calculate the corresponding square error. The first one with the least square error yields the number of nearest sites for the best regional solar radiation estimation. The distance of the farthest site in such a situation corresponds to the radius of influence. As an example, herein, only site-14 calculations are presented in Table 6.2. It is obvious from this table that when Eq. 6.3 is applied by considering the 11 nearest sites the estimation error square becomes minimum with the radius of influence equal to 127.47 km.

Table 6.2 Site 14 adaptive estimation

Number of nearest sites (1)	Estimation (2)	Square error (3)
2	31.94	1.58
3	29.46	13.96
4	30.75	6.00
5	32.42	0.60
6	32.21	0.97
7	32.38	0.67
8	32.82	0.14
9	32.59	0.37
10	32.72	0.23
11	**33.21**	**0.009**
12	33.12	0.01
13	33.48	0.08
14	33.80	0.36
15	33.69	0.24
16	33.63	0.18
17	33.57	0.13
18	33.55	0.12
19	33.48	0.08
20	33.51	0.09

Application of the above adaptive procedure to solar radiation records results in the estimation values, number of the nearest sites with the minimum squared error, and the radius of influence, which are presented in Table 6.3.

It is possible to deduce the following points from this table:

1. The adaptive estimation procedure gives an average solar radiation concentration value similar to average measurements with a 2% error. Hence, it is similar to all the previous methods of adaptive estimation and yields reasonable average values.
2. Comparison of average relative error in Table 6.3 with average relative errors in Table 6.1 shows clearly that the adaptive method with 5.06% error is the best among all approaches and the reduction in the relative error implies that deviations from the average level are taken into account effectively. Figure 6.9 presents the adaptive estimations together with the measured values.
   If Figs 6.8 and 6.9 are compared, it is then obvious that deviations are better accounted for by the adaptive method.
3. The adaptive approach provides the radius of influence for each station as shown in the last column of Table 6.3 The average radius of influence is about 87 km with maximum and minimum values at sites 7 and 2 (and 12), respectively.

By making use of the radius of influence from Table 6.3, it is possible to construct an equal radii regional map as shown in Fig. 6.10. From this map one can know the relevant radius of influence for any desired point within the study area.

6.6 Regional Estimation

Table 6.3 Estimations and radius of influence

Station (1)	Measurement (2)	Adaptive estimation (3)	Relative error (%) (4)	Number of nearest sites (5)	Radius of influence (km) (6)
1	35.50	33.10	6.76	10	199.06
2	29.40	29.10	1.02	2	41.23
3	36.80	36.91	0.29	5	69.46
4	33.70	33.84	0.41	3	69.46
5	35.30	35.83	1.50	3	47.17
6	32.50	32.50	0.00	9	164.50
7	30.60	33.82	9.87	14	222.99
8	30.10	31.63	5.08	3	92.19
9	40.10	37.96	5.34	4	82.46
10	31.60	32.73	3.57	2	67.27
11	34.80	34.77	0.09	4	61.03
12	28.60	30.50	6.64	3	41.23
13	41.50	38.65	7.11	2	58.31
14	33.20	33.21	0.03	11	127.47
15	34.30	34.26	0.17	4	94.34
16	31.00	33.17	7.00	8	127.47
17	29.60	30.14	1.82	3	50.25
18	40.40	39.50	2.23	3	52.20
19	28.50	28.82	1.12	2	51.48
20	24.40	32.34	32.54	3	51.48
21	39.50	33.83	14.35	4	60.21
Average	33.40	33.64	5.06	–	87.20

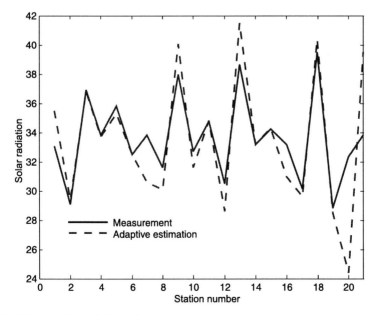

Fig. 6.9 Adaptive solar radiation estimation

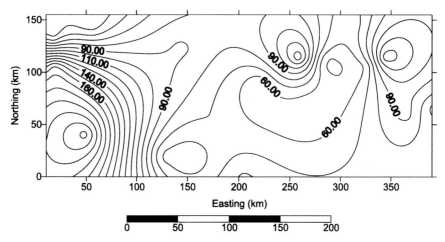

Fig. 6.10 Solar radiation radius of influence map in kilometers

Once, this radius of influence is determined then a circle with the center at the estimation point is drawn. The measurement sites within this circle are taken into consideration in the application of Eq. 6.3 for regional estimation through the SDF weights.

6.6.2 Spatial Interpolation

After having completed the cross-validation procedure and the map of the radius of influence, it is now time to present the spatial interpolation procedure with SDF usage as follows:

1. Select any certain number, say 15, of spatially scattered points within the study area as shown in Fig. 6.3. These sites are locations without measurements. For the sake of argument, they are selected rather arbitrarily with easting and northing coordinates as shown in Table 6.4.
2. The radius of influence of each site is determined from the map in Fig. 6.10 and written in column 4 of Table 6.4.
3. Consideration of the radius of influence for each site defines the number of measurement sites within this radius that are the basis for the solar radiation estimation through Eq. 6.3. Hence, the measurement sites that will be considered in the spatial interpretation of the solar radiation at the site are identified.
4. Subsequently, distances between the interpolation site and the effective measurement sites are calculated.
5. The entry of these distances on the horizontal axis in the SDF (Fig. 6.7b) yields the weights on the vertical axis.
6. Substitution of all the relevant values into Eq. 6.3 provides the solar radiation value estimations at each site, which are shown in the last column of Table 6.4.

6.6 Regional Estimation

Table 6.4 Regional interpolation

Station	Easting (km)	Northing (km)	Radius of influence (km)	Adaptive estimation
(1)	(2)	(3)	(4)	(5)
1	36	136	63	30.02
2	27	90	172	32.71
3	86	116	97	30.13
4	45	29	200	33.81
5	95	47	118	34.88
6	186	23	68	38.49
7	218	58	50	32.63
8	222	129	95	31.77
9	268	98	80	32.35
10	327	134	93	33.47
11	272	43	58	28.99
12	340	40	80	32.30
13	331	65	77	32.78
14	327	87	70	33.08
15	368	96	103	34.37

7. In order to check the reliability of the estimations, the question is now whether these spatial estimations will yield almost the same SDF or not. For this purpose, the SDF calculation steps in Sect. 6.5.1 are applied to the data in Table 6.4.
8. Figure 6.11 indicates the resulting SDF for the measured and estimated solar radiation values. The maximum relative difference between these two SDFs is less than 5% which confirms the practical validity of the SDF adaptive estimation procedure methodology.

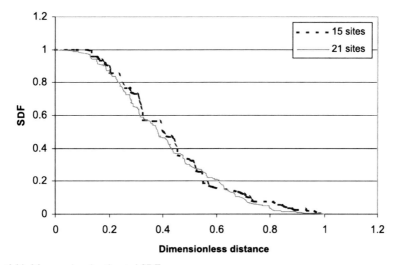

Fig. 6.11 Measured and estimated SDFs

It is now time to compare the estimation procedures by plotting the measurement data versus estimations as in Fig. 6.12 for the global and IDS approaches and for the adaptive technique in Fig. 6.13.

It is obvious that global estimation and inverse distance square approaches cannot pass the test against the measurement because the scatter points do not appear around the 45° straight line, which indicates the best model verification (Chap. 4.3). These models have scatter diagrams with almost horizontal trends which are away from the best model verification line. However, in Fig. 6.13, the SDF model adaptive estimation technique has the trend of scatter points that is very close to the best model verification line.

6.7 General Application

The application of SDF is presented for 29 stations in Turkey as given with their locations in Fig. 4.6 and geographic features in Table 4.3. The recorded monthly average solar radiation amounts are given in Table 6.5.

In order to apply and indicate the reliability of the proposed approaches, stations are considered one by one for cross-validation. Let us say that Ankara is chosen as the estimation site and January as the month of estimation. Figure 6.14 shows the CSV and thereof obtained SDF for January. Herein, on the horizontal axis the dimensionless distances are as the ratios of distances to the maximum distance between Izmir and Van, which is equal to 916.40 km.

Although according to Table 6.5 Ankara has an average January solar radiation record of 5.88 MJ/m, it will be assumed non-existent for the cross-validation.

The subsequent step is to apply the estimation process as explained in the previous section. For this purpose, it is necessary to consider the distances from Ankara to all other 28 stations. Table 6.6 includes these distances in the third column. For the sake of comparison, the fourth and fifth columns include ID and IDS values.

In the sixth column the dimensionless distances are given and they are necessary prerequisites for the SDF in Fig. 6.14b. Dimensionless distances are calculated by dividing each distance value in the third column by the maximum distance value. In the seventh column, the SDF weightings are included as found for January from Fig. 6.14b corresponding to the dimensionless distances.

In the application of SDF for global estimation the available measurement sites are considered in the weighting procedure according to Eq. 6.3 with all stations. The plots of these estimations and the actual measurements are presented in Fig. 6.15.

It is obvious that global estimation procedure appears to be successful on the average. This is tantamount to saying that consideration of all the measurement sites without any distinction causes smoothing in the solar radiation spatial estimation. Relative error percentages of more than 10% appeared excessively at almost all the sites.

In order to improve the situation, it is suggested to use adaptive estimation so that the spatial estimation error becomes minimum. For this purpose, during the

6.7 General Application

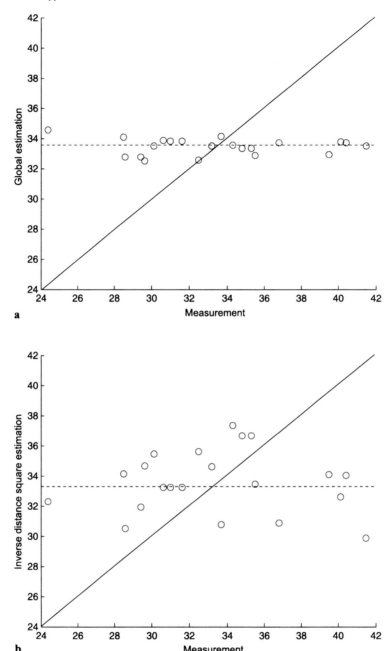

Fig. 6.12 a Global and b IDS estimations

Table 6.5 Monthly average solar radiation values (MJ/m^2)

Station No.	Name	Location Long.	Lat.	Months J	F	M	A	M	J	J	A	S	O	N	D
1	Adana	36.98	35.30	7.68	9.95	13.72	17.57	20.05	21.41	21.88	19.83	16.42	11.99	8.47	6.83
2	Adiyaman	37.75	38.28	6.42	8.73	12.19	15.62	17.75	19.64	19.12	17.16	14.58	10.63	7.18	5.48
3	Afyon	38.75	30.53	7.40	10.46	15.00	18.17	20.55	23.89	23.43	22.22	18.40	12.27	8.01	6.31
4	Amasya	40.65	35.85	5.31	8.26	12.67	16.18	19.01	21.99	21.82	20.10	15.91	10.09	6.09	4.79
5	Anamur	36.10	32.83	7.99	10.80	14.23	18.82	21.58	23.22	22.62	20.52	17.73	12.54	8.85	7.04
6	Ankara	39.95	32.88	5.88	8.81	12.83	15.34	19.15	21.54	21.82	20.52	17.08	11.28	7.37	4.76
7	Antalya	36.88	30.70	9.31	11.93	16.61	20.44	23.25	25.87	25.26	22.40	20.24	14.61	9.94	8.00
8	Aydin	37.85	27.83	8.41	11.06	15.13	19.42	22.59	25.46	25.44	23.38	19.51	14.01	9.01	7.35
9	Balikesir	39.65	27.87	4.78	7.14	10.07	13.67	17.11	19.63	20.01	18.17	14.99	9.76	5.64	3.91
10	Bursa	40.18	29.07	5.96	7.64	10.43	13.37	16.55	19.11	18.87	17.74	14.67	9.96	5.89	4.62
11	Çanakkale	40.13	26.40	6.32	8.95	12.66	16.70	20.44	22.20	22.28	20.37	16.77	11.21	7.22	5.12
12	Çankiri	40.60	33.62	6.35	9.80	14.18	16.31	19.80	21.94	22.02	20.53	17.25	11.77	6.90	4.87
13	Diyarbakir	37.92	40.20	6.02	9.74	14.29	18.68	22.99	26.64	25.67	23.19	19.63	13.90	8.47	6.00
14	Elazig	38.67	39.22	6.18	9.41	13.28	16.58	20.10	23.77	23.70	21.14	17.57	11.67	6.82	5.10
15	Erzincan	39.73	39.50	7.48	10.89	14.44	16.64	18.46	21.94	22.05	19.69	16.70	11.43	7.00	6.23
16	Eskisehir	39.77	30.52	6.38	9.83	14.47	17.04	20.31	22.70	23.02	21.15	17.91	12.12	7.94	5.55
17	Istanbul	40.97	29.08	4.87	7.52	11.12	15.18	19.18	21.80	21.04	18.81	14.83	9.68	6.16	4.51
18	Isparta	37.57	43.77	7.04	9.60	12.87	15.18	17.27	19.33	19.10	17.56	15.14	10.87	7.42	5.81
19	Izmir	37.70	30.55	7.25	10.24	14.16	17.98	21.01	23.98	23.49	21.34	18.52	12.91	8.46	6.57
20	Kars	38.40	27.17	8.46	11.86	16.41	17.80	18.94	22.09	22.83	20.50	17.62	11.91	7.87	6.87
21	Kastamonu	40.60	43.08	5.30	8.21	11.59	13.27	15.94	17.85	18.20	16.97	13.68	8.65	5.79	4.11
22	Kayser	41.37	33.77	6.62	9.29	12.90	15.61	18.50	21.35	22.29	19.60	16.73	11.02	7.23	5.79
23	Kirsehir	38.72	35.48	7.80	11.08	15.20	17.95	20.69	22.94	23.05	21.05	18.01	12.39	8.29	6.74
24	Konya	36.13	34.17	7.28	10.84	14.69	17.75	20.83	23.58	23.73	21.83	17.84	12.74	8.37	6.16
25	Malatya	37.87	32.50	6.60	9.48	13.00	18.14	21.21	24.74	25.17	22.42	18.51	12.18	7.65	5.28
26	Mersin	38.35	38.30	8.62	11.38	16.19	20.02	22.19	24.11	23.78	22.25	19.33	14.26	9.60	7.56
27	Samsun	36.82	34.60	5.61	8.01	10.92	14.27	17.99	20.83	20.52	18.49	14.63	9.54	6.20	4.96
28	Trabzon	41.28	36.33	5.21	7.72	10.78	13.42	15.82	18.18	15.50	14.58	11.78	8.58	5.85	4.76
29	Van	41.10	39.72	9.29	12.93	16.40	19.91	22.39	25.22	24.28	22.39	19.16	13.21	9.53	7.48

6.7 General Application

Table 6.6 Distances between Ankara and other stations and weightings for January

Station No.	Name	Distance (km)	ID (10³) (1/km)	IDS (10⁶) (1/km²)	Dimensionless distance*	SDF value	Estimation (MJ/m²) CSV	Estimation (MJ/m²) ID	Estimation (MJ/m²) IDS
1	Adana	391.7	2.553	6.518	0.427	0.730	5.60	19.60	50.04
2	Adiyaman	527.6	1.895	3.592	0.576	0.466	2.99	12.17	23.07
3	Afyon	242.1	4.130	17.057	0.264	0.922	6.82	30.57	126.24
4	Amasya	263.6	3.794	14.391	0.288	0.909	4.82	20.14	76.40
5	Anamur	428.1	2.336	5.456	0.467	0.646	5.16	18.66	43.59
6	**Ankara**	**0.0**	**0.000**	**0.000**	**0.000**	**0.000**	**0.00**	**0.00**	**0.00**
7	Antalya	390.6	2.560	6.554	0.426	0.733	6.82	23.84	61.03
8	Aydin	495.4	2.019	4.075	0.541	0.534	4.49	16.97	34.26
9	Balikesir	429.2	2.330	5.427	0.468	0.640	3.06	11.14	25.96
10	Bursa	325.2	3.075	9.455	0.355	0.836	4.98	18.33	56.38
11	Çanakkale	551.9	1.812	3.283	0.602	0.452	2.86	11.44	20.74
12	Çankiri	95.7	10.446	109.115	0.104	0.996	6.32	66.33	692.89
13	Diyarbakir	671.9	1.488	2.215	0.733	0.260	1.57	8.96	13.33
14	Elazig	563.6	1.774	3.148	0.615	0.443	2.74	10.97	19.47
15	Erzincan	565.6	1.768	3.126	0.617	0.427	3.19	13.23	23.39
16	Eskisehir	202.4	4.940	24.407	0.221	0.951	6.06	31.50	155.60
17	Istanbul	340.9	2.934	8.606	0.372	0.818	3.98	14.28	41.90
18	Isparta	315.3	3.171	10.056	0.344	0.854	6.01	22.31	70.74
19	Izmir	521.4	1.918	3.678	0.569	0.467	3.38	13.90	26.66
20	Kars	867.9	1.152	1.328	0.947	0.053	0.45	9.75	11.24
21	Kastamonu	174.8	5.720	32.719	0.191	0.972	5.15	30.31	173.35
22	Kayser	262.1	3.815	14.556	0.286	0.910	6.03	25.27	96.41
23	Kirsehir	231.0	4.329	18.738	0.252	0.929	7.24	33.77	146.20
24	Konya	233.6	4.281	18.324	0.255	0.927	6.74	31.14	133.31
25	Malatya	500.0	2.000	4.000	0.546	0.533	3.52	13.19	26.39
26	Mersin	378.9	2.639	6.964	0.414	0.754	6.50	22.74	60.02
27	Samsun	326.6	3.062	9.377	0.356	0.831	4.66	17.17	52.59
28	Trabzon	590.1	1.695	2.872	0.644	0.376	1.95	8.82	14.95
29	Van	916.4	1.091	1.191	1.000	0.000	0.000	10.14	11.07

* Dimensionless distance = distance/ 916.40

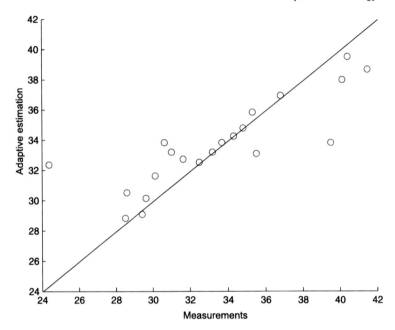

Fig. 6.13 Adaptive estimation

estimation procedure at a particular site, the number of adjacent stations is increased from one (the nearest site) to the total number of sites in order of increasing distance. Consequently, it is observed that each site has its special number of adjacent sites for the best interpolation depending on the regional variability of the solar radiation with the minimum error.

The consequent estimates resulting from the ID, IDS, and SDF are shown in Table 6.7 where the number of adjacent sites is also presented.

In the same table comparison of relative errors from these three approaches indicates that the SDF has the least values – almost all of which fall within the acceptable limit of 10%. Last but not the least, Fig. 6.16 shows the variation in the measured and adaptively estimated values of solar radiation.

Even the visual comparison of Figures 6.15 and 6.16 shows clearly that the adaptive spatial solar radiation estimation leads to great improvements. A practical question is how many solar radiation stations should be considered in the spatial estimation at any given unmeasured site? In order to provide an objective answer, it is necessary to provide equal adjacent site contours for the whole region of the study area. Figure 6.17 indicates such a map and the following features can be depicted:

1. In general the eastern Mediterranean and the south-eastern parts of Turkey require the least adjacent stations for the solar radiation spatial adaptive estimation. In fact, in the south-eastern parts two measurement stations are sufficient for estimation. The regional climatology of Turkey indicates that these regions have less rainfall, long sunshine duration hours, and relatively high temperatures.

6.7 General Application

Table 6.7 Actual and estimated solar radiation values

Station No.	Name	Actual	No. of adjacent sites	Estimation Adjacent sites CSV	All sites CSV	ID	IDS	Relative error (Adjacent sites, CSV)	Relative error (All sites, CSV)	Relative error (ID)	Relative error (IDS)
1	Adana	7.68	3	7.69	6.89	7.16	7.87	0.21	10.21	6.75	2.42
2	Adıyaman	6.42	2	6.39	6.88	6.77	6.63	0.48	6.66	5.11	3.10
3	Afyon	7.40	5	7.18	6.75	6.78	6.79	2.95	8.80	8.34	8.20
4	Amasya	5.31	3	5.75	6.63	6.56	6.13	7.69	19.92	19.11	13.46
5	Anamur	7.99	2	7.95	6.98	7.08	7.48	0.50	12.60	11.33	6.34
6	Ankara	5.88	3	5.83	6.70	6.69	6.53	0.90	12.24	12.04	9.91
7	Antalya	9.31	5	7.61	6.82	6.90	7.07	18.28	26.75	25.94	24.03
8	Aydin	8.41	5	7.14	6.72	6.77	6.89	15.10	20.12	19.52	18.00
9	Balikesir	4.78	4	6.00	6.83	6.75	6.58	20.27	29.95	29.14	27.29
10	Bursa	5.96	5	5.93	6.71	6.44	5.87	0.50	11.11	7.41	1.58
11	Çanakkale	6.32	6	6.25	6.65	6.56	6.14	1.07	5.00	3.73	2.75
12	Çankiri	6.35	7	6.10	6.59	6.45	5.99	3.93	3.64	1.62	5.69
13	Diyarbakir	6.02	2	6.30	6.95	6.85	6.73	4.51	13.39	12.14	10.52
14	Elazig	6.18	2	6.31	6.86	6.77	6.67	1.97	9.79	8.63	7.29
15	Erzincan	7.48	7	6.52	6.71	6.58	6.30	12.87	10.38	12.01	15.78
16	Eskisehir	6.38	5	6.23	6.83	6.73	6.63	2.28	6.66	5.27	3.77
17	Istanbul	4.87	2	5.38	6.71	6.60	6.24	9.51	27.48	26.18	21.99
18	Isparta	7.04	8	7.52	6.86	7.12	7.62	6.43	2.43	1.21	7.71
19	Izmir	7.25	6	6.65	6.76	6.86	7.21	8.28	6.78	5.37	0.59
20	Kars	8.46	3	7.36	6.40	6.58	6.39	13.00	24.35	22.29	24.45
21	Kastamonu	5.30	4	5.79	6.70	6.69	6.48	8.55	20.97	20.78	18.18
22	Kayser	6.62	8	6.85	6.74	6.84	6.95	3.31	1.67	3.12	4.64
23	Kirsehir	7.80	4	7.55	6.77	6.90	7.08	3.28	13.18	11.53	9.26
24	Konya	7.28	2	7.42	6.84	7.01	7.27	1.95	5.93	3.63	0.04
25	Malatya	6.60	5	6.54	6.82	6.69	6.49	0.82	3.31	1.41	1.64
26	Mersin	8.62	3	7.82	6.86	7.03	7.42	9.24	20.38	18.37	13.85
27	Samsun	5.61	2	5.30	6.55	6.49	5.91	5.44	14.33	13.52	5.11
28	Trabzon	5.21	3	5.55	6.80	6.86	6.92	6.22	23.41	24.15	24.82
29	Van	8.29	2	7.27	6.71	6.76	6.88	12.36	19.05	18.43	17.07

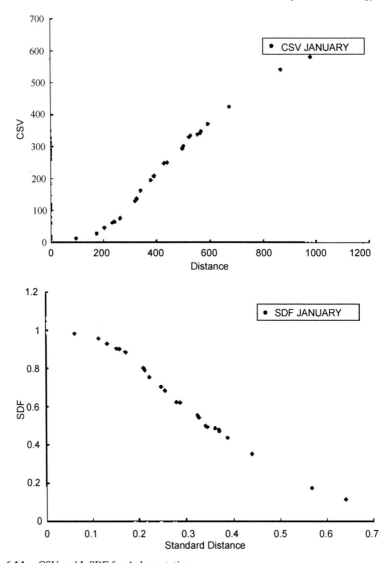

Fig. 6.14 a CSV and b SDF for Ankara station

2. In the southwestern part, up to five adjacent site numbers are necessary. This is well correlated with the topographic heights in this region, which include several lakes that effect the evaporation and rainfall regimes.
3. The adjacent site number increases toward the eastern border where there are the very rugged and high elevation mountain chains of Turkey. Winters are long, about 5–7 months each year.

6.7 General Application

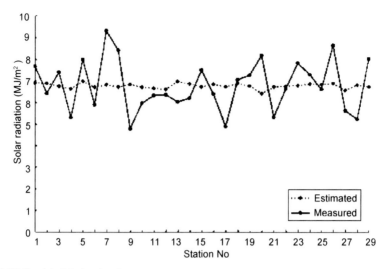

Fig. 6.15 Spatial global estimation

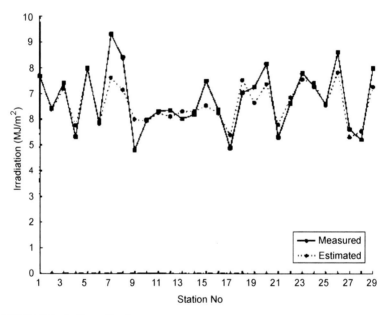

Fig. 6.16 Spatial adaptive estimation

4. Another high adjacent site requirement appears in the north along the middle Black Sea coast where the elevations reach up to almost 3000 m. Severe winter conditions also occur in this region.

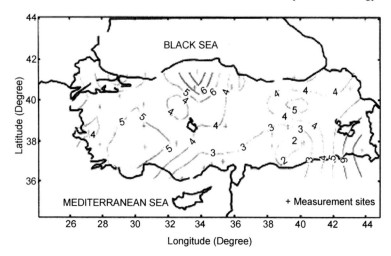

Fig. 6.17 Equal adjacent site contours

References

Bardossy A, Bogardi I, Kelly WE (1990a) Kriging with imprecise (fuzzy) variograms. I: theory: Math. Geology 22:63–79
Bardossy A, Bogardi I, Kelly WE (1990b) Kriging with imprecise (fuzzy) variograms. I: application: Math. Geology 22:81–94
Barnes SL (1964) A technique for maximizing details in numerical weather map analysis. J Meteorol 3:396–409
Box GEP, Jenkins GM (1970) Time series analysis forecasting and control. Holden Day, San Francisco
Clark I (1979) Practical geostatistics. Applied Science Publishers, London
Cressie NAL (1993) Statistics for Spatial Data. Wiley, New York
Cressman GP (1959) An operational objective analysis system, Monthly Weather Rev 87:367–374
Davis J (1986) Statistic and data analysis in geology. Wiley, New York
Delhomme JP (1978) Kriging in the hydrosciences. Adv Water Resour 1:251
Dooley JE, Hay JE (1983) Structure of the global solar radiation field in Canada. Report to the Atmospheric Environment Service, Downsview, contact no. DSS-39SS-KM601.0.1101, 2 vol
Gandin LS (1963) Objective analysis of meteorological fields (translated from Russian by the Israel Programme for Scientific Translations, Jerusalem). Hydrometeorological Publishing, Leningrad
Hay JE (1979) Calculation of monthly mean solar radiation for horizontal and inclined surfaces. Solar Energy 23:301–307
Hay JE (1983) Solar energy system design. The impact of mesoscale variations in solar radiation. Atmos Ocean 24:138
Hay JE (1984) Errors associated with the spatial interpolation of mean solar irradiance. Solar Energy 37:135
Hay JE (1986) Errors associated with the spatial interpolation of mean solar irradiance. Solar Energy 37:135
Journel AG, Huijbregts CJ (1989) Mining Geostatistics. Academic, New York
Krige DG (1951) A statistical approach to some basic mine evaluation problems on the Witwateround. J Chimic Min Soc S Afr 52:119–139
Matheron G (1965) Les Variables Regionalisees et leur Estimation. Masson, Paris

References

Matheron G (1971) The theory of regionalized variables and its applications. Ecole de Mines, Fontainbleau

Perrie W, Toulany B (1965) Correlation of sea level pressure field for objective analysis. Monthly Weather Rev 117(9):1965–1974

Sasaki Y (1990) An objective analysis for determining initial conditions for the primitive equations. Technical Rep (ref 60-167) (College station: Texas A and M University)

Şen Z (1989) Cumulative semivariogram model of regionalized variables. Int J Math Geol 21:891

Şen Z (1991) Standard cumulative semivariograms of stationary stochastic processes and regional correlation. Int J Math Geol 24:417–435

Şen Z (1997) Objective analysis by cumulative semivariogram technique and its application in Turkey. J Appl Meteorol 36:1712–1724

Şen Z (1998) Average areal precipitation by percentage weighting polygon method. J Hydrol Eng 1:69–72

Şen Z, Habib Z (2001) Monthly spatial rainfall correlation functions and interpretations for Turkey. Hydrol Sci J 46:525–535

Şen Z, Şahin AD (2001) Spatial interpolation and estimation of solar irradiation by cumulative semivariogram. Solar Energy 71:11–21

Suckling PW, Hay JE (1978) On the use of synoptic weather map typing to define solar radiation regimes. Monthly Weather Rev 106:1521–1531

Thiebaux HS, Pedder MA (1987) Spatial objective analysis, Academic, New York

Thiessen AD (1912) Precipitation averages for mountainous areas. Monthly Weather Rev 39:1082–1084

Zelenka A (1985) Satellite versus ground observation based model for global irradiation. INTERSOL 58, proceedings of the 9th biennial congress ISES, vol 4. Pergamon

Zelenka A, Czeplak G, D'Agostino V, Weine J, Maxwell E, Perez R (1992) Techniques for supplementing solar radiation network data, vol 2. A report IEA task 9. Report no. IEA-SHCP-9D-1

Chapter 7
Solar Radiation Devices and Collectors

7.1 General

The practical applications and beneficial uses of solar radiation require consideration of engineering aspects in order that the use of the solar energy is efficient and sustainable. For instance, in any design of solar energy powered devices, it is necessary to know how the power density will vary during the day, from season to season, and also the effect of tilting a collector surface at some angle to the horizontal. From the practical point of view, for most purposes solar energy applications can be divided into two components, namely, direct (beam) radiation and scattered or diffuse radiation, as already mentioned in Chap. 3.

Solar energy is expected to be the foundation of a sustainable energy economy, because sunlight is the most abundant renewable energy resource. Additionally, solar energy can be harnessed in an almost infinite variety of ways beginning with simple solar cookers now used in different parts of the world. There is a vast amount of literature about the use of solar energy both in engineering and architectural design procedures and projects (Leng 2000). It is not possible to present all these studies herein, but the proper references with a brief description will be provided in this chapter so that the reader can find further information on the topic.

7.2 Solar Energy Alternatives

The nuclear fusion reactions in the sun yield a huge amount of energy which is estimated at 3.47×10^{24} kJ per unit time. Only a small part, about 5×10^{-11} kJ, of this energy is irradiated onto the earth's surface. The incident solar energy is distributed in many ways as shown in Fig. 7.1. Solar energy is clean, undepletable, and harmless to living organisms on the earth because the harmful short wavelength ultraviolet rays are absorbed before reaching the troposphere by the stratospheric

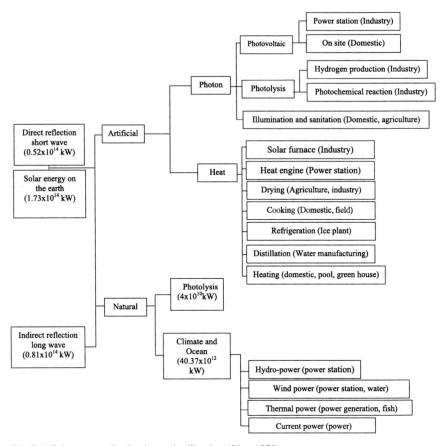

Fig. 7.1 Solar energy distribution and utilization (Ohta 1979)

ozone layers and weakened by the air composition and moisture in the troposphere (Şen 2004; Chap. 2 and 3). Solar energy energizes the atmosphere and thus generates climatic phenomena, but the balance of the energy is absorbed by molecules of the materials on the earth and converted into heat at low temperatures. This is an example of the entropy increasing process of nature. It is, therefore, necessary to plan actively to utilize the sun's photon and high temperature heat energies before they decay to produce entropy. The artificial utilization of solar energy is also shown in Fig. 7.1. Two classifications, namely, natural and artificial are apparent from this figure and the photon energy is of better quality and is much higher than that of the heat energy.

In fact, all energy sources with the exception of atomic energy have a solar energy origin. A sweeping statement yet true to the extent that even coal, oil, and natural gas are forms of solar energy. In order to separate the various forms of solar energy, the following three categories are adopted:

1. Heat from the sun's rays which is possible when there is little or no cloud cover. This type of energy is dependent on heat from the sun's rays and is dominated by the multiplicity of methods designed to heat water.
2. Power from the sun's light any time except at night, cloudy or clear.
3. Power from air or water movement (hydrological cycle), any time day or night, cloudy or clear.

7.3 Heat Transfer and Losses

The easiest way of using collected solar radiation is for *low temperature* heating purposes. Most of the low-temperature solar heating systems depend on the use of glazing, because it has the ability to transmit visible light and to block infrared radiation. *High-temperature* solar collectors employ mirrors and lenses. Solar thermal engines are an extension of *active solar heating* and help to produce high temperatures to drive steam turbines to produce electric power. Solar ponds and even ocean thermal energy conversion devices that operate on the solar-induced temperature difference between the top and the bottom of the world's oceans may cover many hectares.

Another way of benefiting from solar radiation is by *passive solar heating* devices which have different meanings. For instance, in the narrow sense, it means the absorption of solar energy directly into a building to reduce the energy required for heating the habitable space. Passive solar heating systems are integral parts of the building and mostly use air to circulate the collected energy without pumps or fans. In the broad sense, passive solar heating means low-energy building designs, which are effective in reducing the heat demand to the point where small passive solar gains make a significant contribution in winter.

It is well known that black surfaces absorb solar radiation more than any other color and, therefore, when a surface is blackened it will absorb most of the incident solar radiation. Continuous flow of solar radiation onto such a surface will increase its temperature. This will continue until the heat gain from the solar radiation is in equilibrium with the heat loss from the collector. Of course, among the heat losses, there are two types, namely, naturally unavoidable losses and losses due to human uses. The heat can be transmitted to where it is needed through pipes soldered to the metal plate which is heated due to exposure to solar radiation. The heat balance of a collector will have three components in relation as follows (ASHRAE 1981; Dunn 1986):

Absorbed heat – Lost heat = Removed heat by coolant

It is possible to define the *coefficient of efficiency* for the collector as

Efficiency coefficient = (absorbed heat – lost heat)/incident solar radiation.

In practice, the collectors must be designed in such a manner that the efficiency becomes high. In order to achieve such a goal there are two methods, either the reduction of heat losses or the increase of the incident solar radiation and, hence, the heat absorbed per unit area. For low-temperature collectors heat loss reduction methodology is suitable. It is possible to reduce heat loss by using transparent cover plates, by using specially treated absorber surfaces, and by evacuating the space between the cover plate and the absorber surface. In contrast, for high-temperature solar collectors the efficiency must be increased by increasing the incident radiation through the concentrators. Of course, for this purpose only direct radiation is considered.

There are three heat transfers that should be considered in any solar energy design for efficiency. For any solar radiation collector to work efficiently it is necessary to reduce the heat losses or to minimize them. As a material is heated by solar radiation, it seeks to reach equilibrium with its surroundings by conduction, convection, and radiation processes.

7.3.1 Conduction

This corresponds to heat transfer within a solid body where there are at least two different heat areas, *i. e.*, a temperature difference. Such a heat transfer is possible by means of vibrations of the atomic lattice which forms the body of the material. The heat is also carried away by electrons, and this contribution is much greater than that due to lattice vibration. During *conduction* there is no mass transfer. Atoms move randomly under thermal stress in liquids and gases, and they also lead to heat conduction. The heat transfer is proportional to the temperature difference, dT, along a distance, dx, (dT/dx being the *temperature gradient*) and, hence, the *conduction heat flow* can be expressed as

$$H_c = -k\frac{dT}{dx}, \qquad (7.1)$$

where H_c is the heat flow per unit area of cross-section (W/m^2), T is the temperature (°C), x is the direction and distance (m), and finally, k is the *thermal conductivity* of the material (W/m°C). Thermal conductivity is special for each material and its value is given for various materials in Table 7.1.

As solar radiation is absorbed by an opaque material, the energy redistributes itself as it is conducted between adjacent molecules. Such redistribution is dependent, on the one hand, on temperature difference and, on the other, on the thermal conductivity of the material. Metals, in general, have big conductivities, and consequently, can transmit large amounts of energy under small temperature gradients. However, in insulators the reverse situation is valid where under large temperature gradients only a small amount of heat is conducted. It is known that air is a very good insulator. Hence, most of the practical insulators rely on very small pockets of air trapped between the panels of glazing, as bubbles in a plastic medium, or between the fibers of mineral wools.

7.3 Heat Transfer and Losses

Table 7.1 Thermal conductivity of some materials

Material	k (W/m°C)
Metals:	
Copper	385
Aluminum	205
Steel	50
Non-metals:	
Glass	0.8
Concrete	0.8
Wood	0.14
Saw-dust	0.06
Rock wool	0.04
Polystyrene (expanded)	0.03
Glass fiber	0.03
Liquids:	
Water	0.61
Gases:	
Hydrogen	0.142
Helium	0.142
Air	0.0239

7.3.2 Convection

This is the process by which heat from the hot surfaces is carried away by a fluid such as water. Fluid flowing across a surface is heated and then the heated volume is removed due to fluid flow with replacement by new, cold fluid. This heat transfer is referred to as *convective* cooling or heating. The rate of heat removal will depend on both the temperature difference between the surface and the bulk fluid temperature, and also on the velocity and characteristics of the fluid. Another sort of convective heat transfer can be considered for a horizontal hot plate in still air where the air adjacent to the top surface will become hotter than the bulk of the air. As a result of hot air expansion and density decrease, hot air is replaced by cooler air. In solar energy conversion both forced and natural convections may be accompanied by phase changes. Hence, *convective heat flow* can be expressed as

$$H_f = c\left(T_s - T_f\right) = c\Delta T , \qquad (7.2)$$

where H_f is the heat flow per unit area (W/m^2), c is the convective *heat transfer coefficient* (W/m^2°C), T_s is the surface temperature (°C), and T_f is the fluid temperature (°C). The actual calculation of c is somewhat complicated, because it is dependent on both the nature of the fluid and also on its flow velocity. Approximate convective heat transfer coefficients are given for flat plate collectors in Table 7.2.

This refers to the transference of heat to a fluid (gas or liquid). First, energy is transferred to the molecules of the fluid, which then physically move away, taking the energy with them. A warmed fluid expands and rises, as a result, creating

Table 7.2 Convective heat transfer coefficient (Dunn 1986)

System configuration	c(W/m^2°C)
Heat transfer between parallel plates (separation 2.5 – 10 cm) with air at atmospheric pressure	3
Heat transfer from the surface of a cover plate, where v is the wind velocity at the surface of the plate in meters per second	$2.3 + 3.8v$

a current known as natural *convection*, which is one of the principal processes of heat transfer through windows. It occurs between the air and glass. It is possible to reduce the convection losses through double glazing windows by filling the space between the glazing with heavier, less mobile gas molecules, such as argon or carbon dioxide. On the other hand, since the convection currents cannot flow in a vacuum, the space between the glazing may be evacuated.

7.3.3 Radiation

A hot body may lose heat by radiation through emission of EM waves (Chap. 3). The maximum power which can be radiated from a body at a given temperature is called the black body radiation corresponding to that temperature. The radiation power, P, from a *black body* increases as the fourth power of the *absolute temperature*, T, of the same body and it is given by *Stefan's law* as

$$P = \sigma T^4, \tag{7.3}$$

where $\sigma = 5.67 \times 10^{-8}$ W/m^{2}K^4 is *Stefan's constant*. Heat flux in the case of radiation from a black body is presented in Table 7.3 for different absolute temperatures.

Table 7.3 Black body radiation

Surface temperature (K)	Heat flux (W/m^2)
6000	73.5×10^6
3000	4.6×10^6
2000	9.1×10^5
1073	75.0×10^3
873	32.9×10^3
673	11.6×10^3
473	2.84×10^3
353	1.10×10^3
333	880
300	459

Table 7.4 U-values of different types of window construction

Window type	U-value (W/m²°C)
Single-glazed window	6
Double-glazed window	3
Double-glazed window with "low-E" coating	1.8
Double-glazed window with heavy gas filling	1.5
Experimental evacuated double-glazed window with transparent insulation spacers	0.5
For comparison: 10-cm opaque fiberglass insulation	0.4

Similar to the sun's radiation, heat can be radiated from the surfaces of heated materials. The amount of radiation is first dependent on the temperature of the radiating body and then on the destination of the radiation. In low-heat solar collectors on roofs, energy radiates to the atmosphere. The amount of radiation is also dependent on the surface material *emissivity*. Most materials used in building construction have high emissivity of approximately 0.9, which means that they radiate 90% of the theoretical maximum for a given temperature. Usually, the total heat loss from the combined effects of conduction, convection, and radiation is referred to as the *U-value*. Its unit is the amount of loss per area per degree centigrade. Typical U-values are provided in Table 7.4.

7.4 Collectors

Typical uses of solar radiation *collectors* can be grouped into four different categories depending on the purpose:

1. As a low-temperature heat source which may be used for domestic hot water or crop drying purposes.
2. In order to power heat engines, relatively high heat collectors can be used.
3. Depending on the climate, the collectors can be used as high temperature heat to power refrigerators and air conditioners.
4. Photovoltaic (PV) cells are used for direct electricity production.

Most low-temperature solar collectors are dependent on the properties of glass which is transparent to visible light and short-wave infrared, but opaque to long-wave infrared reradiated from a solar collector or building behind it. In order to benefit from daylight and, especially, solar radiation as the energy source, manufacturers strive to make glass as transparent as possible by keeping the iron content down. In Table 7.5 the optical properties of some commonly used glazing materials are indicated.

In order to design a solar energy powered device, it is necessary to know how the power density will vary during the day and seasonally at the site concerned. It is also important to consider the tilting of the collector surface with the horizontal. As

Table 7.5 Optical properties of commonly used glazing materials

Material	Thickness (mm)	Solar transmittance	Long-wave infrared transmittance
Float glass (normal window glass)	3.9	0.83	0.02
Low-iron glass	3.2	0.90	0.02
Perspex	3.1	0.82	0.02
Polyvinyl fluoride	0.1	0.92	0.02
Polyester	0.1	0.89	0.18

has already been explained in Chap. 3, it is possible to consider the solar radiation in two parts, namely, direct radiation and scattered or diffuse radiation. In solar engineering device designs, most often the direct radiation amounts are significant. The relative proportions of direct to diffuse radiation depend on the day of the year, meteorological conditions, and the surrounding site. The diffuse component on a clear day is usually not more than 20% depending on the circumstances. On the other hand, on an overcast day this proportion may become almost 100%.

7.4.1 Flat Plate Collectors

The *flat plate collectors* are based on two important principles: a black base that absorbs the solar radiation better than any other color and a glass lid that is needed to keep the heat in. Figure 7.2 shows the cross-section of the most commonly used flat plate collector. Its surface should be located perpendicularly to the solar radiation direction for the maximum solar energy gain.

Here the sun's rays go through the glass cover and the air layer to warm up the black metal plate which in turn warms the water. Unfortunately, the ordinary metal plate is also warmed up. The heat insulation lagging keeps most of the heat inside the sandwich. With the heat in the water, it has now to be moved to where good use can be made of it. The simplest method for achieving this water movement is shown in Fig. 7.2, the *"thermo-siphon"* system. Its operation is based on the simple fact that hot water will rise to settle above a quantity of cooler water.

As the collector heats up, the water in it rises out at the upper pipe and pushes its way into the top of the tank. This hot water then displaces some of the cold in the tank, pushing it down and out of the bottom. This heat-induced circulation is completed as the water, being pushed down the pipe, comes round the bottom and back into the collector.

Different types of solar collectors are given in Fig. 7.3. Among these, the most primitive is unglazed panels which are most suitable for swimming pool heating where it is not necessary for the collectors to raise the temperature of the water to more than a few degrees above ambient air temperature, so heat losses are relatively unimportant.

7.4 Collectors

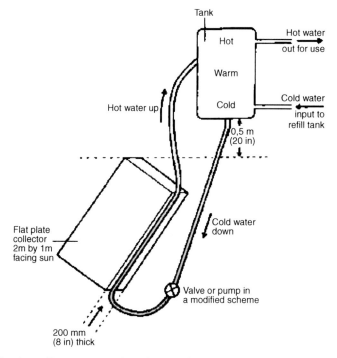

Fig. 7.2 Flat plate collector cross-section (Howell 1986)

Flat plate collectors are the main stay of domestic solar water heating. These are usually single glazed, but may have an additional second glazing layer. The more elaborate the glazing system, the higher the temperature difference that can be sustained between the absorber and the external wall. It is necessary and is usual that the absorber plate should have a black surface with high absorptivity. In general, most black paints reflect approximately 10% of the incident radiation. On the other hand, flat plate air collectors are mainly used for space heating only. These type of collectors are connected with photovoltaic panels for producing both heat and electricity. Evacuated tube collectors in Fig. 7.3 are in the form of a set of modular tubes similar to fluorescent lamps. The absorber plate is a metal strip down the center of each tube. A vacuum in the tube suppresses convective heat losses.

In practice, most often the collectors do not move, and therefore, they must be located such that during one day the maximum amount of solar radiation can be converted into solar energy. For this reasons, fixed collectors must be located to face south (north) in the northern (southern) hemisphere. This implies that for given a latitude there is a certain angle which yields the maximum solar energy over the year. As a practical rule, for low latitudes the angle of the collector is almost equivalent to the angle of latitude, but increases by 10° at 40°N and 40°S latitudes. All these arrangements are for flat-surfaced collectors. Typical temperatures that can be achieved by flat plate collectors vary between 40°C and 80°C depending on the as-

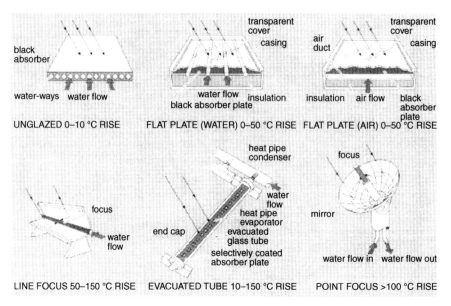

Fig. 7.3 Various collectors

tronomic, topographic, and meteorological conditions. In a flat plate collector, the energy incident on the surface cannot be increased and all that can be done is to ensure that surface absorbs as much as possible of the incident radiation, and that the losses from this surface are reduced as far as possible. Figure 7.4 shows a flat plate collector.

Some of the incident radiation is lost by reflection but for a blackened surface about 95% of the radiation will be absorbed. The heat losses from flat plate collectors are shown in the same figure. In general, in these collectors the lower surface usually has an insulating layer of material such as several centimeters of glass wool. The heat can be lost through the conduction, convection, and radiation mechanisms.

Flat plate collectors are usually roof mounted and their tracking of the sun is not possible. They are subject to many external events such as frost, wind, sea spray, acid rain, and hail stones. They must also be resistant against corrosion and significant temperature changes. Low-temperature flat plate collectors are able to raise the water temperature up to boiling point in the summer, provided that they are double-glazed and the water circulation is not fast enough to carry away the heat quickly. These may be only a few square meters in area. In order to collect enough solar energy to supply the winter demand, the collectors would have to cover a large area and in such cases the solar energy production during the summer would not be wholly exploited. This means a wastage of the capital expenditure.

7.4 Collectors

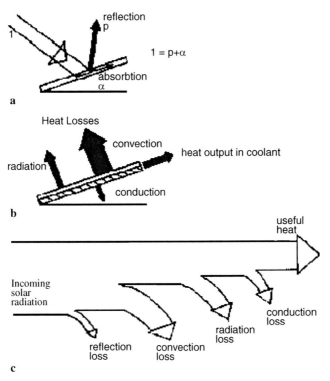

Fig. 7.4 a–c. Flat plate solar collectors (Dunn 1986)

7.4.2 Tracking Collectors

Logically, in order to collect the maximum radiation for each unit surface area of collector, it is necessary to direct the collector surface at right angles to the direction of direct radiation. Continuation of maximum benefit by the collector during a day is possible by keeping the collector surface perpendicular to the incident solar radiation throughout the hours of daylight. It is, theoretically, simple to show that a *tracking* (moving) *collector* compared to the horizontal collector at the same site will collect $\pi/2$ times more energy per day. However, in practice this factor is around 1.5 times. Of course, the more the direct radiation, the better is the energy generation from the sun's radiation.

7.4.3 Focusing (Concentrating) Collectors

If high temperatures are needed then the collector surface is manufactured as a curve for *focusing* (concentrating) the solar radiation at certain points by a mirror or lens. Mirrors are cheaper to construct than lenses. The *mirror collectors* may have spherical parabolic or linear parabolic shapes as shown in Fig. 7.5. In a parabolic mirror solar radiation is concentrated at a point and, therefore, the *concentration ratio* is approximately 40,000 whereas the concentration for a one-dimensional device of a linear parabolic system is around 200.

So far as the *lens collectors* are concerned there are single surface or equivalent Fresnel types as shown in Fig. 7.6. Although in flat plate collectors diffuse solar radiation also makes a contribution in the radiation collection, the concentrating collectors focus the incident sunlight on the collector surface, leaving the contribution of diffuse radiation aside.

Another disadvantage of the concentrating collectors is that they must track the sun in order to obtain the optimal benefit. Concentrating collectors rise the temperature of the heater up to 300–6000 °C. These collectors must be aligned with sufficient accuracy to ensure that the focus coincides with the collector surface. The greater the degree of concentration, the more accurate is the alignment required.

Line focus collectors concentrate the solar radiation onto a pipe running down the center of a trough, and are mainly used for generating steam for electricity generation. To get the maximum benefit, it is necessary that the trough is pivoted to track the sun's movement in any direction. *Point focus collectors* as shown in Fig. 7.7 also track the sun but in two dimensions and these also generate steam for conversion into electricity.

If the solar radiation is concentrated through mirrors or lenses then temperatures in excess of the boiling point of water may be reached. It is possible to use such high temperatures through steam production for mechanical work, for instance, for water

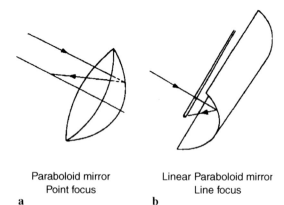

Fig. 7.5 a,b. Parabolic mirror concentrator of solar radiation

a Paraboloid mirror
   Point focus

b Linear Paraboloid mirror
   Line focus

7.4 Collectors

Fig. 7.6 a–d. Single surface and equivalent Fresnel lenses

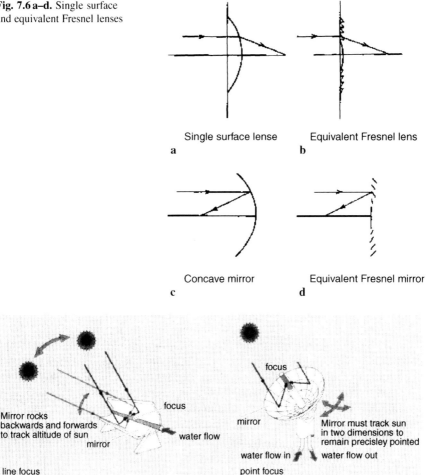

Fig. 7.7 Line and point focus by moving mirrors

pumping or electricity generation. These are named *high-temperature collectors*. Most often parabolic mirrors are used for solar radiation concentrations. As shown in Fig. 7.7 all the sun's rays directed parallel to the axis of such a mirror will be reflected to one point.

It is necessary that the mirror tracks the sun, otherwise slightly off-axis solar beams will make inconvenient reflections, and the intensity of the radiation concentration onto a point or line will be weakened. In the line focus form the sun's radiation can be concentrated on a small region running along the length of the mirror. For the maximum focusing of the sun's radiation, it is necessary to tract the sun in an elevation that is only up and down. However, in the point focus form, the sun's radiation is reflected on a boiler in the mirror center. For optimum performance, the axis must be pointed directly at the sun at all times, so it needs to track the sun both in elevation and in azimuth (Chap. 3).

Another technology of centralized electricity generation is *solar-thermal power*. These are produced by using large mirrored troughs to reflect the sun's rays onto an oil-filled tube, which in turn superheats water to produce the steam that drives an electricity-generating turbine. Since the mid-1980s, about 350 MW of these solar energy systems have been installed across three square miles of the southern Californian desert and these are enough to supply electricity to 170,000 homes. Especially, in areas of extensive *pollution control*, solar-thermal electricity substitution is required urgently for the pollution reduction. In order to produce sufficient energy, solar-thermal electricity production is only practical in areas where there are intense direct sunlight conditions such as the arid regions of the world. Another way of harnessing solar power is to use an array of mirrors to concentrate, or focus, sunlight onto water flowing through a metal pipe. The resulting steam can then be used to drive a turbine.

7.4.4 Tilted Collectors

It is necessary to have *tilted* surfaces for the maximum collection of solar energy. The angle of tilting is dependent both on the latitude and the day of the year. If the tilt angle is equal to the latitude then the sun's rays will be perpendicular to the surface of collector at midday in March and September. For the maximization of solar collection in the summer it is convenient to tilt the surface a little more toward the horizontal. However, for the maximization in the winter the surface must be tilted more to the vertical. Fortunately, the effects of tilt and orientation are not particularly critical. Table 7.6 presents totals of energy incident on various tilted surfaces in the London region.

Table 7.6 Collector surface tilting to the south at 53° N latitude near London

Tilt (°)	Annual total radiation (kWh/m^2)	June total radiation (kWh/m^2)	December total radiation (kWh/m^2)
0 - horizontal	944	155	16
30	1068	153	25
45	1053	143	29
60	990	126	30
90 - vertical	745	82	29

7.4 Collectors

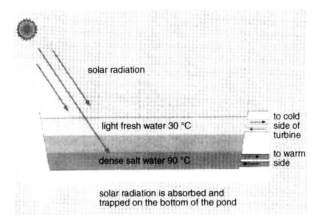

Fig. 7.8 Solar pond

7.4.5 Solar Pond Collectors

Instead of flat plate collectors, *solar ponds* are used for thermal electricity production. A solar pond is similar to a large salty lake as shown in Fig. 7.8. In such a pond, salty water is required to be at the bottom with fresh water at the top.

The difference in salt concentration causes a gradient in the salt concentration. The incident solar radiation is absorbed directly from the sun's radiation in the bottom of the pond. The hot and salty water cannot rise, because it is heavier than the fresh water in the top layer. Hence, the upper fresh water layer acts as an insulating blanket and the temperature at the bottom of the lake can reach 90 °C.

The thermodynamic limitations of the relatively low solar-to-electricity conversion efficiencies are typically less than 2%. One of the main advantages of the solar pond system is that the large thermal mass of the pond acts as a heat store, and electricity generation can go on day or night, as required. However, a large amount of fresh water is needed to keep the solar pond running with a proper salt gradient.

7.4.6 Photo-Optical Collectors

Among the *photo-optical transmission* methods, Çinar (1995) has considered the collection of radiation by focused collectors. On the other hand, Baojun *et al.* (1995) have investigated the solar energy relationships with *fiber-optic radiation*.

The most advanced and recent method of solar energy collection, as well as transmission, is by using fiber-optics. Collection of energy directly as light by concentrator collectors causes almost no energy loss in the transmission. Since the collected solar radiation is in the form of light, it can be used directly for lighting purposes. However, it can also be used for heating and for conversion into electrical energy, if desired.

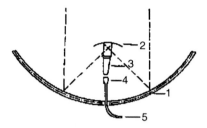

Fig. 7.9 Fiber optical collector system. *1* Large diameter convex collector, *2* small diameter convex reflector, *3* refining lens system, *4* lens system that renders the solar rays into parallel form, *5* fiber optic glass transmission cable

After collection of the solar energy through focusing, it is refined by means of a lens system and, finally, directed toward a fiber-optic glass transmission cable. The transmission is affected without any further loss to the desired area over long distances as shown in Fig. 7.9. It is obvious that large diameter convex collectors collect the incoming radiation, and then send it to another small diameter convex reflector.

The small dish in Fig. 7.9 reflects the incoming radiation to the refining lens system. This system refines the radiation twice after the focusing. The light ray that is refined down to the size of a needle goes through a collector which includes a set of lenses that render the radiation into a parallel beam. Such a condensed solar ray enters without any loss into fiber-optical cable which has a high transmission capability.

Through the aforementioned system, the transmission of solar energy will be possible, without losses, from solar radiation rich regions of the world to solar radiation poor regions. For this purpose, a regional energy transmission network must be constructed. In this manner, the solar energy can be transmitted to consuming countries where the solar radiation possibilities are rather poor. For instance, when the central European and Arabian conditions are considered, because of the low solar potential of the central Europe, the solar energy transmission from Arabian deserts is possible through the above-mentioned system. Figure 7.10 includes the fiber-optic glass cable transmission system among the selected regions of the Arabian and northern African desert regions to European countries.

The significance of this topic can be appreciated from the solar energy figures presented in Table 7.7 concerning central Europe and the Arabian Rub-Al-Khali desert, which covers about $660,000 \text{ km}^2$ and from each square meter of which 1 kW/h solar energy can be generated.

The solar energy collection area is about $360 \times 10^9 \text{ m}^2$ and, hence, $360 \times 10^9 \text{ m}^2 \times 1 \text{ kWh} = 360 \times 10^9 \text{ kWh} = 360 \times 10^9 \text{ MW/h}$ solar energy can be harvested which is equal to $1440 \times 10^9 \text{ MW/year}$. By considering about 6 m^2 of surface area for each collector, it is possible to find the number of necessary collectors to be $360 \times 10^9 / 6 = 60 \times 10^9$.

Due to the location and planning of some housing complexes, lighting problems might exist and such undesirable situations can be avoided by including fiber-optic

7.4 Collectors

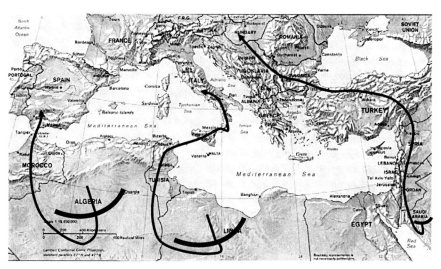

Fig. 7.10 Fiber-optic collector energy transmission

Table 7.7 Average solar energy per square kilometer in central Europe and Arabia

Region	Total annual sunshine duration (h)	Radiation (kcal/h)	Radiation (kcal/year)	Energy (kWh)	Energy (kW/year)
Central Europe	1200	172	206,200	0.2	240
North-eastern Turkey	1825	344	627,800	0.4	730
Arabian deserts	4000	860	3,440,000	1.0	4000

systems in the architectural designs. Such a system may even provide facilities for multi-story greenhouse activities (Chap. 2). By leading the solar radiation over fruit, vegetables, and flowers in multi-story buildings, a covered agricultural production area may be established, and consequently, cheap and healthy food production may become available. The application of fiber-optic electrical energy production in the future is expected to minimize the demand on fossil fuel energy and to provide continuity in renewable energy availability, especially by exploiting the abundantly existing solar radiation potential in the world. The collection of solar energy through fiber-optic glass and lens systems causes insignificant losses and transmission takes place instantaneously.

In particular, the transmission of solar radiation to regions with very little variation provides opportunity for its use directly as light in heating and electrical and hydrogen-generation purposes. Photo-optical energy plants are a means for using solar radiation at low cost, and hence the demand on fossil fuels such as coal, petroleum, and natural gas will decrease leading to a clean atmospheric environment. The energy obtained in this manner may also be used for electrolysis of water into hydrogen and oxygen leading to hydrogen energy production. This will increase the efficiency of solar-hydrogen energy prospects and future usages.

7.5 Photovoltaic (PV) Cells

PVs or *solar cells* (SC) convert sunlight directly into electricity. When photons strike certain semiconductor materials, such as silicon, they dislodge electrons which causes a potential difference to form between the specially treated front surface of the SCs and the back surface. In order to increase the voltage, individual cells are combined in a panel form. The most advanced photon utilization technology is the SC to which the PV effects of semiconductors are applied. SCs are the standard bearer of the new energy technologies because of their great potential. Their successful development is dependent on cost reduction of the power-generating systems that include SCs. They must either be used together with storage devices or as supplements to conventional facilities. Due to their high cost they are still not practical for large-scale power generation. The few central solar generation facilities in operation are experimental and need large areas of land. With current technology about $10\,m^2$ of PV panels are required to generate 1 kW of electricity in bright sunlight. It would take hundreds of square kilometers of *solar panels* to replace an average nuclear power plant. For instance, $220{,}000\,km^2$ would be needed to supply the world with power. Some scientists suggest that the size of the solar power footprint could be reduced by as much as 75% by placing satellites in space to collect sunlight, convert it into electricity, and then beam the power to the earth's surface in the form of microwaves. Currently, the problems for researchers in SC technology are making SCs more reasonable in price and more efficient. Unfortunately, high efficiency and low cost tend to be mutually exclusive.

Photovoltaic cells consist of a junction between two thin layers (positive, p, and negative, n) of dissimilar semiconducting materials. When a photon of light is absorbed by a valance electron of an atom, the energy of the electron is increased by the amount of energy of the photon. If the energy of the photon is equal to or more than the band gap of the semiconductor, the electron with the excess energy will jump into the conduction band where it can move freely. However, if the electron does not have sufficient energy to jump into the conduction band, the excess energy of the electron is converted to excess kinetic energy of the electron, which manifests as an increase in temperature. If the absorbed photon has more energy than the band gap, the excess energy over the band gap simply increases the kinetic energy of the electron. One photon can free up only one electron even if the photon energy is greater than the gap band. Figure 7.11 indicates schematically a PV device.

As free electrons are generated in the n layer by the photon action they can either pass through an external circuit or recombine with positive holes in the lateral direction, or move toward the p-type semiconductor. However, the negative charges in the p-type semiconductor at the p-n junction restrict their movement in that direction. If the n-type semiconductor is made extremely thin, the movement of electrons and therefore the probability of recombination within the n-type semiconductor are greatly reduced unless the external circuit is open. In this case the electrons recombine with the holes and an increase in the temperature of the device is observed. The energy of a photon is already expressed by Eq. 3.3 and by considering the light

7.5 Photovoltaic (PV) Cells

speed from Eq. 3.1 the photon energy can be obtained as

$$E_p = \frac{hc}{\lambda} \tag{7.4}$$

Photovoltaic cells are usually manufactured from silicon although other materials can also be used. n-type semiconductors are made of crystalline silicon that has been "doped" with tiny quantities of an impurity (usually phosphorous) in such a way that the doped material possesses a surplus of free electrons. On the other hand, p-type semiconductors are also made from crystalline silicon, but they are doped with very small amounts of a different impurity (usually boron) which causes the material to have a deficit of free electrons. Combination of these two dissimilar semiconductors produces an n-p junction, which sets up an electric field in the region of the junction (Fig. 7.11). Such a set up will cause negatively (positively) charged particles to move in one direction (in the opposite direction).

Light is composed of a steam of tiny energy particles called photons, and if photons of a suitable wavelength fall within the p-n junction, then they can transfer their energy to some of the electrons in the material so prompting them to a higher energy level. When the p-n junction is formed, some of the electrons in the immediate vicinity of the junction are attracted from the n-type layer to combine with holes on the nearby p-type layer. Similarly, holes on the p-type layer near the junction are attracted to combine with electrons on the nearby n-type layer. Hence, the net effect is to set up around the junction a layer on the n-type semiconductor that has more positive charges than it would otherwise have.

In recent years, power generation from renewable resources has been counted upon to bridge the gap between global demand and supply of power. The direct conversion technology based on solar PV devices has several positive attributes and

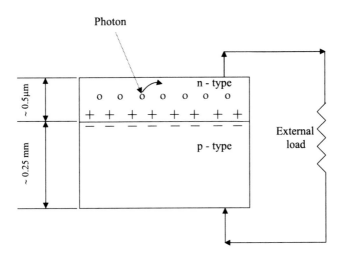

Fig. 7.11 Simple PV cell and resistive load

seems to be most promising. Extensive research activities over the past 25 years have led to significant cost reduction and efficiency amelioration (De Meo and Steitz 1990; Kaushika 1999).

Generation of electricity from sunlight started in the 1950s when the first PV cell was invented, which converted solar radiation directly into electric current via a complex photo-electric process. PV technology has advanced during the last five decades, making it possible to convert a larger share of sunlight into electricity; it has reached as much as 14% in the most advanced prototype systems. Although the cost of PV devices has fallen drastically during recent decades, it is still four to six times the cost of power generation from fossil fuels. PV devices are already the most economical way of delivering power to homes far from utility lines. It is expected that this technology will become an economic way of providing supplementary utility power in rural areas, where the distance from plants tends to cause a voltage reduction that is otherwise costly to remedy. As they become more versatile and compact, PV panels could be used as roofing material on individual homes, bringing about the ultimate decentralization of power generation. For instance, the desert areas are the most attractive and rich regions of the world for the solar radiation conversion into electric power. One day in the future the world's deserts may become very large solar power plants, which may centralize power in the same way as do today's coal and nuclear power plants. PV panels are much more effective in hazy or partly cloudy conditions and they can be installed even on very small scale residential rooftops.

Photovoltaic solar cells are semiconductor diodes that are designed to absorb sunlight and convert it into electricity. The absorption of sunlight creates free minority carriers, which determine the solar cell current. These carriers are collected and separated by the junction of the diode, which determines the voltage. PV SCs have been the power supply of choice for satellites since 1958. Light drives the PV process and provides the energy that is converted into electricity. PV cells use primarily visible radiation. The distribution of color within light is important because a PV cell produces different amounts of current depending on the various colors shining on it. Infrared radiation contributes to the production of electricity from crystalline silicon and some other materials. In most cases infrared radiation is not as important as the visible portion of the solar spectrum (Chap. 3).

Terrestrial applications of PV devices developed rather slowly. Some of the main advantages of their use as an electric power source can be given as follows (Deniz 2006, unpublished):

1. Direct conversion of solar radiation into electricity
2. No mechanical moving parts and no noise
3. No high temperatures
4. No pollution
5. PV modules are very robust and have a long life
6. The energy source (sun) is free and inexhaustible
7. PV energy is a very flexible source; its power ranges from microwatts to megawatts

7.6 Fuel Cells

A *fuel cell* is an electrochemical energy converter. It converts the chemical energy of fuel (H_2) directly into electricity. A fuel cell is like a battery but with constant fuel and oxidant supply (Fig. 7.12).

Fuel cells are preferred for the following reasons (Barbir 2005):

1. Promise of high efficiency
2. Promise of low or zero emissions
3. Run on hydrogen/fuel may be produced from indigenous sources/issue of national security
4. Simple/promise of low cost
5. No moving parts/promise of long life
6. Modular
7. Quiet

7.7 Hydrogen Storage and Transport

It is an unfortunate characteristic of solar energy that it arrives in a quite random manner depending on the meteorological conditions and it does not arrive at all time to suit our needs. Since the time of usage does not always match with the time of availability, it is necessary to store the solar energy at times of availability so as to use it at times of need.

The need for new and renewable energy alternatives due to the depletion of conservative energy sources also brought about studies on the efficient usage and transmission of available energies. As is well known, the major criticism against these energy alternatives is the problem of energy storage (Tsur and Zemel 1992). Uneven solar energy potential in the world causes an imbalance in its production among various regions, some of which are relatively richer in solar energy than others. Such imbalances can be avoided only through an efficient energy transportation system.

If the *storage* and *transmission* of solar energy can be achieved then the coal, fuel oil, and natural gas requirements of any country will be reduced significantly. Such solar energy transmission system will provide benefits for great trade centers,

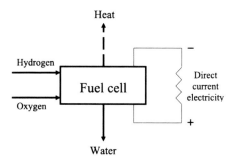

Fig. 7.12 Fuel cell principle

factories, and, especially, its application to illuminate green plants will lead to reduction in the fossil energy use to a minimum, and provide continuity in the renewable energy alternatives.

Any discrepancy between the energy supply and demand can be offset by hydrogen storage and its use at the time of need as a source of energy. Hydrogen can be stored on a large scale underground in the aquifers, in depleted petroleum or natural gas reservoirs, and in artificial caverns as a result of mining activities. The latter method is the most commonly used alternative in some countries. Hydrogen can be transported to the places of consumption from the production plants in gaseous form through underground pipelines and by supertankers in liquid form. Hydrogen can be stored in stationary or mobile storage systems at the consumer site depending on the end use. It can be stored either as a pressurized gas or as a liquid, or using some of its unique physical and chemical properties, in metal hydrides and in activated carbon. Hydrogen can be used instead of fossil fuels virtually for all purposes as a fuel for surface and air transportation, heat production, and electricity directly (in fuel cells) or indirectly (through gas and steam turbine driven generators) (Veziroğlu 1995).

Hydrogen can be converted to electricity electrochemically in fuel cells with high efficiency. It is not subject to Carnot cycle limitations, which is the case with the present day thermal power plants whether they burn fossil or nuclear fuels. It has been stated by Veziroğlu (1995) that Tokyo Electric Utility started experimenting with a 4.5-MW United Technologies fuel cell years ago. Now, they have another 11-MW fuel cell on line.

Another unique property of hydrogen is that it will combine with certain metals and alloys easily, in large amounts, forming hydrides in exothermic chemical reactions. Hydrogen is released when the hydrides are heated. The temperature and pressure characteristics vary for different metals and alloys. Many household appliances working with hydrogen do not need CFCs and, hence, they will not damage the ozone layer.

Hydrogen has the further property that it is flameless when it burns or the catalytic combustion is in the presence of small amounts of catalysts, such as platinum or palladium. Catalytic combustion appliances are safer, have higher second thermodynamic law efficiencies, and are environmentally compatible.

The "technology readiness" of hydrogen energy systems needs to be accelerated, particularly in addressing the lack of efficient, affordable production processes; lightweight, small volume, and affordable storage devices; and cost-competitive fuel cells. The hydrogen energy system has the potential to solve two major energy challenges that confront the world today: reducing dependence on petroleum imports and reducing pollution and greenhouse gas emissions.

7.8 Solar Energy Home

Careful building design makes the best use of natural daylight. In order to make the best use of solar energy, it is necessary to understand the climate of the region. Buildings that are inappropriate for the local climate cause energy wastage (Howell 1986). In order to gather radiation directly by devices, house roofs are constructed as discrete solar collectors.

It is possible to consider a south-facing window as a kind of passive solar heating element. Solar radiation will enter during daylight hours, and if the building's internal temperature is higher than that outside then heat will be conducted and convected back out. Here, the main question is whether more heat flows in than out, so that the window provides a net energy benefit. The answer depends on the following several points:

1. The internal temperature of the building
2. The average external air temperature
3. The available amount of solar energy
4. The transmitting characteristics, orientation, and shading of the window
5. The U-value (see Sect. 7.3.3) of the window whether it is single or double glazed

The total amount of heat needed for supply over the year can be called the gross heating demand, which can be supplied from three sources:

1. The body heat of people and heat from cooking, washing, lighting, and appliances are together named as "free heat gains" in a house or apartment. Although, individually, they are not significant, collectively they may amount to 15 kWh/day. Free heat gains help in reducing the space heat loading.
2. Passive solar gains occur mainly through the windows.
3. Fossil fuel energy exploitation from the normal heating system.

If the house is insulated properly, it is not necessary to have large areal collectors, because the energy need will be small. Here lies the key problem in active solar space heating: either to insulate the house to have less energy demand or to build poorly insulated houses and try to implement solar energy for space heating.

7.9 Solar Energy and Desalination Plants

Water is an extremely important commodity for the improvement of arid (desert) and semi-arid environments. As for the water production technology, *desalination plants* widely use fossil fuels. Hence, for the improvement of these regions it is necessary to shift from fossil fuel usage to some environmentally friendly energy source, such as solar energy as it is available abundantly in such environments. It is necessary to develop a sustainable water production system using the renewable energy that is presented by solar energy instead of fossil fuels in these regions. Specifically, the Arabian Gulf countries have the latest water production technology and the use of the solar energy alternative for this purpose must be investigated in spite of fossil fuel availability. The relationships of the natural energy sources and the sea water desalination technology are shown in Fig. 7.13.

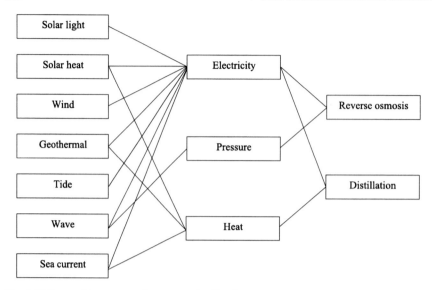

Fig. 7.13 Renewable energy and sea water desalination

At present, the energy sources used for water production are mainly heat and pressure. The latter is produced from electricity and heat or pressure can be made from various renewable energy sources as shown in Fig. 7.13. Among them solar energy is the most universal and exists in abundance especially in desert environments. Currently, there is a problem in the solar energy usage cost performance. However, the expenses after equipment construction are almost zero and the water production cost becomes low.

7.10 Future Expectations

In general, there are two main reasons for future energy research. First, as a result of global warming, atmospheric and environmental pollution due to energy consumption, present day energy patterns, using predominantly fossil fuels, must be either improved in their quality or more significantly, they must be substituted with more environmentally reliable clean and renewable energy sources. The second reason for future research is the appreciation that the fossil fuel reserves are limited and bound to be exhausted sooner or later. If the necessary precautions are not taken from now on by radical innovations in energy systems and their technologies, then future human generations on the earth will face an extremely precarious position. Additionally, population increase places extra pressure on the energy resources and the energy consumption per capita per day in developing countries is about 10 oil-equivalent-liter, which is below one-tenth of that in industrial countries (Chap. 1). In order to produce new energy sources independent of fossil and nuclear fuels, the following points must be considered in future research programs:

1. The solar beam collector with a Fresnel lens or concave mirror
2. Electric charge separation by solar radiation
3. Other natural processes that reduce entropy, such as the functions of a membrane, catalyst, biological organ, other chemical phenomena, *etc.*

In the long run, full consideration must be given to the amount of energy that is required to produce more energy. One of the constant research areas is storage and the two most promising new devices are *silica gel beds* and *two-vessel storage* (Ohta 1979). Silica gel beds try to improve the efficiency of pebble storages. It is possible to obtain the same performance with a volume fifteen times less. The silica gel beds are relatively unaffected by thermal losses so there is also a saving on insulation. On the other hand, the two-vessel store introduces a fresh storage technique. As Howell (1986) explained, the idea relies on the chemical reaction that occurs when acid and water are mixed; heat is then released. Hence, for heat storage it can be used to drive water and acid into separate vessels where they can remain for years as stored energy. By allowing the acid back into the water the stored heat is released.

It is necessary all over the world to reduce the cost of solar collectors although this may appear in the guise of increased efficiency at the same cost. This is tantamount to saying that as production increases and the days of handmade collectors pass, the labor content of the product will reduce to a minimum. As the only other major production cost is the cost of materials, the other move must be toward cheaper materials.

Although copper and aluminum make excellent devices to heat water, as collector material one must not forget that they are only intermediaries. The objective is to heat fluid not metal. Therefore, future research on solar collectors is into the use of plastics, and many more alternatives might follow which combine the advantages of suitability, mass production, cheap raw materials, and long life. Replacement of glass with a layer of clear fluorescent tubes reduces the cost almost fivefold.

It is expected that within the next two decades solar energy, whether transmitted through electrical lines or used to produce hydrogen, will become the cornerstone in the global energy policy. In the future, wherever solar energy is abundant, hydrogen can be produced without pollution and shipped to distant markets. For this purpose, the Sahara Desert in Africa can be regarded as the solar-hydrogen production area from where the hydrogen can be transmitted to consumption centers in Europe. Germany leads the effort to develop solar-hydrogen systems. There are demonstration electrolysis projects powered by PV cells already operating in Germany and the solar energy rich deserts of the Kingdom of Saudi Arabia. Germany spends some $25 million annually on hydrogen research projects.

The invention of optical fibers has led to extensive studies on the traditional methods of illumination and sterilization using the sun's radiation. Optic fibers provide a pathway to transmit solar beams almost anywhere. Çınar (1995) has explained such transmission of solar energy from sunshine-rich desert areas to exploitation

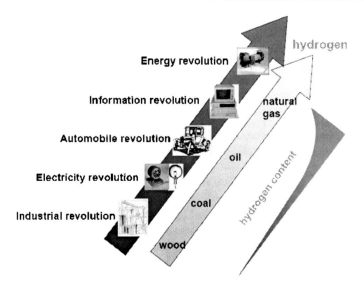

Fig. 7.14 Evolution of modern civilization (Barbir 2005, unpublished)

centers. The solar radiation incident on the Fresnel lenses is focused at a point where the entropy of the system is greatly reduced. If the temperature of the point of focus is 300 °C and the ambient temperature is 27 °C, then the entropy of the focus is reduced by about half. Searching for similar entropy-reducing natural phenomena is an important task in energy science. The application fields of solar energy are well known and rather traditional, but new technologies will have an impact and will eventually be put to practical use.

In the two past centuries there were many revolutions that propelled society into a new mode of development and the majority of these revolutions are energy related as shown in Fig. 7.14. It seems that in the future energy-related revolutions are going to take place in addition to stress on water resources, which might be relieved through use of the practically inexhaustible solar energy supply and desalination plant production of additional water for the survival of humanity. Hydrogen energy is also related to water production in this respect.

References

ASHRAE (1981) Handbook of fundamentals, chap 27. American Society of Heating Refrigerating and Air Conditioning Engineers, New York

Baojun L, Dong W, Zhou M, Xu H (1995) Influences of optical fiber bend on solar energy optical fiber lighting. 2nd international conference on new energy systems and conversions, İstanbul, p 41

Barbir F (2005) Electricity generation with fuel cells. UNIDO, International Centre for Hydrogen Energy Technologies. This material is provided to the attendees of the International Advanced

Course on Renewable Energies, at RERDEC İstanbul, June 5–14, and intended for their use only

Çınar MA (1995) Solar heater with thermal energy reservoir. 2nd international conference on new energy systems and conversions, İstanbul, p 457

De Meo EA, Steitz P (1990) In: Boer KW (ed) Advances in solar and wind energy, chap 1. American Solar Energy Society and Plenum, New York

Deniz Y (2006) This material is provided to the attendees of the International Advanced Course on Renewable Energies, at RERDEC İstanbul, March 5–14, and intended for their use only

Dunn PD (1986) Renewable energies: sources, conversion and application. Peregrinus, New York

Howell D (1986) Your solar energy home. Pergamon, New York

Kaushika ND (1999) Design and development of fault-tolerant circuitry to improve the reliability of solar PV modules and arrays. Final technical report of the Department of Science and Technology, the Government of India. Research Project no. III 5(98)/95-ET

Leng G (2000) RETScreen International: a decision-support and capacity-building tool for assessing potential renewable energy projects. UNEP Ind Environ 3, 22–23 July–September

Ohta T (1979) Solar-hydrogen energy systems. Pergamon, New York

Şen Z (2004) Solar energy in progress and future research trends. Progr Energy Combustion Sci Int Rev J 30:367–416

Tsur Y, Zemel A (1992) Stochastic energy demand and the stabilization value of energy storage. Nuclear Resour Model 6:435

Veziroğlu TN (1995) International Centre for Hydrogen Energy Technologies. Feasibility Study. Clean Energy Research Institute, University of Miami, Coral Gables

Appendix A
A Simple Explanation of Beta Distribution

If a random variable x is equally likely to take any value in the interval 0 to 1, then its probability distribution function (PDF) is constant over this range. The beta PDF is a very flexible function for use in describing empirical data such as $\overline{a'}$ and $\overline{b'}$ as in Chap. 4. The general form of this distribution is given by Benjamin and Cornell (1970) as

$$f(x) = \frac{1}{\beta} x^{r-1} (1-x)^{r-\gamma-1} \quad (0 < \gamma < 1),$$

where β is the normalizing constant as

$$\beta = \frac{(r-1)!(t-r-1)!}{(t-1)}$$

for integer r and $t-r$ values, otherwise

$$\beta = \frac{\Gamma(r)\Gamma(t-r)}{\Gamma(t)}$$

in which $\Gamma(.)$ is the incomplete gamma function of the argument. Herein, r and t are the PDF parameters related to the mean $\overline{x}$ and variance σ_x^2 parameters as follows:

$$\overline{x} = \frac{r}{t}$$

and

$$\sigma_x^2 = \frac{r(t-r)}{t^2(t+r)}.$$

Appendix B
A Simple Power Model

The non-linear least squares technique depends on the minimization of the prediction error square summation from a non-linear equation. The non-linearity exists in the power term of the solar radiation model as presented in the text by Eq. 5.12. In order to predict the solar radiation amount $\overline{(H/H_0)}$ at any time instant, say i, from the fractal exponent model there is an error, e_i, involved as follows:

$$\overline{\left(\frac{H}{H_0}\right)}_i = a_p + b_p \overline{\left(\frac{S}{S_0}\right)}_i^{\frac{1}{p}} + e_i \quad \text{(B.1)}$$

or the error term is calculated as

$$e_i = \overline{\left(\frac{H}{H_0}\right)}_i - a_p - b_p \overline{\left(\frac{S}{S_0}\right)}_i^{\frac{1}{p}}$$

and the sum of error squares for n predictions becomes notationally as

$$SS = \sum_{i=1}^{n} e_i^2 = \sum_{i=1}^{n} \left[\overline{\left(\frac{H}{H_0}\right)}_i - a_p - b_p \overline{\left(\frac{S}{S_0}\right)}_i^{\frac{1}{p}} \right]^2 \quad \text{(B.2)}$$

The partial derivatives of this expression with respect to model parameters a, b, and c leads to

$$\frac{\partial SS}{\partial a} = -2 \sum_{i=1}^{n} \left[\overline{\left(\frac{H}{H_0}\right)}_i - a_p - b_p \overline{\left(\frac{S}{S_0}\right)}_i^{\frac{1}{p}} \right] (-1), \quad \text{(B.3)}$$

$$\frac{\partial SS}{\partial b} = -2 \sum_{i=1}^{n} \left[\overline{\left(\frac{H}{H_0}\right)}_i - a_p - b_p \overline{\left(\frac{S}{S_0}\right)}_i^{\frac{1}{p}} \right] \left[-\overline{\left(\frac{S}{S_0}\right)}_i^{\frac{1}{p}} \right], \quad \text{(B.4)}$$

and

$$\frac{\partial SS}{\partial c} = -2\sum_{i=1}^{n}\left[\overline{\left(\frac{H}{H_0}\right)_i} - a_p - b_p\overline{\left(\frac{S}{S_0}\right)_i^{\frac{1}{p}}}\right]\left[-b_p\overline{\left(\frac{S}{S_0}\right)_i^{\frac{1}{p}}\log\left(\frac{1}{p}\right)}\right]. \quad (B.5)$$

In order to find the optimum solution of parameter estimates these three differentials must be set equal to zero:

$$-2\sum_{i=1}^{n}\left[\overline{\left(\frac{H}{H_0}\right)_i} - a_p - b_p\overline{\left(\frac{S}{S_0}\right)_i^{\frac{1}{p}}}\right](-1) = 0, \quad (B.6)$$

$$-2\sum_{i=1}^{n}\left[\overline{\left(\frac{H}{H_0}\right)_i} - a_p - b_p\overline{\left(\frac{S}{S_0}\right)_i^{\frac{1}{p}}}\right]\left[-\overline{\left(\frac{S}{S_0}\right)_i^{\frac{1}{p}}}\right] = 0, \quad (B.7)$$

and

$$-2\sum_{i=1}^{n}\left[\overline{\left(\frac{H}{H_0}\right)_i} - a_p - b_p\overline{\left(\frac{S}{S_0}\right)_i^{\frac{1}{p}}}\right]\left[-b_p\overline{\left(\frac{S}{S_0}\right)_i^{\frac{1}{p}}\log\left(\frac{1}{p}\right)}\right] = 0. \quad (B.8)$$

Hence, there are three unknowns and three equations. However, the analytical and simultaneous solution of these three equations is not possible, and therefore, the numerical solution is sought. For this purpose, first of all it is possible to obtain from Eqs. B.6 and B.7 by elimination the following parameter estimations:

$$b_p = \frac{\frac{1}{n}\sum_{i=1}^{n}\overline{\left(\frac{H}{H_0}\right)_i\left(\frac{S}{S_0}\right)_i^{\frac{1}{p}}} - \left[\frac{1}{n}\sum_{i=1}^{n}\overline{\left(\frac{H}{H_0}\right)_i}\right]\left[\frac{1}{n}\sum_{i=1}^{n}\overline{\left(\frac{S}{S_0}\right)_i^{\frac{1}{p}}}\right]}{\left[\frac{1}{n}\sum_{i=1}^{n}\overline{\left(\frac{H}{H_0}\right)_i^{\frac{1}{p}}}\right] - \left[\frac{1}{n}\sum_{i=1}^{n}\overline{\left(\frac{H}{H_0}\right)_i^{\frac{1}{p}}}\right]^2} \quad (B.9)$$

and

$$a_p = \frac{1}{n}\sum_{i=1}^{n}\overline{\left(\frac{H}{H_0}\right)_i} - b_p\frac{1}{n}\sum_{i=1}^{n}\overline{\left(\frac{S}{S_0}\right)_i^{\frac{1}{p}}}. \quad (B.10)$$

These are the two basic equations that reduce to the linear regression line coefficient estimations for $p = 1$. This situation is equivalent with the AM parameter estimation. The third equation of the non-linear least squares technique can be ob-

B A Simple Power Model

tained from Eq. B.8 as

$$\sum_{i=1}^{n} \overline{\left(\frac{H}{H_0}\right)_i \left(\frac{S}{S_0}\right)_i^{\frac{1}{p}}} - a_p \sum_{i=1}^{n} \overline{\left(\frac{S}{S_0}\right)_i} - b_p \sum_{i=1}^{n} \overline{\left(\frac{S}{S_0}\right)_i^{\frac{1}{p}}} = 0 . \tag{B.11}$$

The numerical solution algorithm is explained in the main text.

Index

A

absolute temperature   244
acid rain   5, 13, 15, 36, 42, 248
actinography   102
adaptation   5, 8, 10, 12, 13, 16, 18, 33, 39, 90
aerosol   26, 33, 62, 65, 70, 169, 209
air   21, 30, 32, 40, 48, 62, 202, 240, 243, 245–247, 260
   conditioning   8, 10
   mass   65, 70, 164, 174
   pollution   4, 14, 22, 29, 36, 38
   quality   15, 29, 118, 177
albedo   23, 24, 63, 115
   planetary   23
   surface   23
Angström model   102, 115, 116
apparent time   74
aspect angle   83
atmospheric   21, 23, 24, 26, 47
   environment   13
   pollution   3, 6, 9, 36
   transmissivity   89, 151
   transmittance   70, 71
   turbidity   117, 151
azimuth angle   68, 75, 77, 104
   error   104

B

beam irradiation   115
beam radiation   61, 70, 71, 76, 87, 91, 116
best linear unbiased estimate   213
bio-mass   3, 4, 6, 35–37, 39, 40, 48
black body   53, 244

C

carbon dioxide cycle   28
chlorofluorocarbon   14, 31
chromosphere   48
clean energy   13, 42
clear sky   61, 64, 70, 93, 97, 106, 114–116, 132, 159
clearness   172
   index   105, 110, 114
climate   1, 3, 5, 12, 22, 23, 26, 47, 71
   change   4, 15, 22, 26, 35, 55
   impacts   44
   variability   38, 211
   vulnerability   18
   wave   40
climatic risks   13
coal equivalent ton   43, 51
coefficient   109, 110
   determination   109, 110
   efficiency   241
collectors   202, 239, 245
   flat plate   243, 246, 249
   focusing   250, 254
   high-temperature   241, 242
   low-temperature   242
   photo-optical   253
   tilted   252
   tracking   249
conditional distributions   118, 119
conduction   34, 242, 245, 248, 256
consumption   9
convection   48, 242–245, 248
convective   48
   heat flow   243
   zone   48
correlation   108, 113, 118, 119, 203

coefficient 110, 127, 210
matrix 135
regional 218
spatial 211, 212, 216
cosine effect error 104
cross-validation 209, 213, 221–223, 226, 228

D

daily insolation 88, 94, 96–98
day-length 152
declination 76–78, 81, 88, 92, 94, 96, 140
deficit 24, 35, 173, 257
defuzzification 185, 186
diffuse 56, 61, 62, 68, 102, 118, 177, 202
  irradiation 101
  radiation 57, 61, 67, 68, 87, 115, 156, 163, 239, 246
  ratio 196
direct 177
direct radiation 57, 61, 87, 88, 113, 177, 242, 246, 249
distance 209, 214
double
  sunrise 93, 97, 98
  sunset 93, 97, 98
doubling time 6, 8
drought 3, 6, 13, 172, 175, 176
  extreme 173
  mild 173
  moderate 173
  severe 173

E

eccentricity 59, 65, 69, 72, 77, 88, 89
economic growth 6, 12, 13, 17
electrolysis 41, 42, 255, 263
emissivity 245
energy
  budget 24
  consumption 3, 4, 6, 8, 10–12, 14, 16, 36, 40, 262
  crisis 5, 18
  demand 4, 6, 7, 9–11, 14, 17, 34–36, 261
  electromagnetic 49, 50, 63
  flux 68
  production 4, 24, 102, 167, 248, 255
  self-sufficiency 11
  spectrum 53
  sustainable 10
environmental 6, 9
  hazards 9, 16
  impacts 14, 16
  management 16, 17
  planning 16
  pollution 10, 42, 262
environmentally friendly 3, 14, 35, 36, 42, 261
equation of time 73
equatorial plane 76, 77, 80
equinoxes 75, 76, 78
error variance 213
extraterrestrial irradiation 88, 91, 116, 121, 160

F

fiber-optic radiation 253
flood 3, 9, 11, 16, 17
  protection 39
fossil fuel 3, 5, 8–11, 14, 15, 28, 29, 34–38, 40, 41, 47, 48, 255, 258, 260–262
fuel cell 259, 260
fusion 34, 49, 50, 239
fuzzy 151
  inference 177
  logic 151
  model 176
  rule base 176, 182, 184
  sets 176, 178, 180

G

gas 240
genetic algorithm 151, 178, 186, 187
geostatistics 211
global
  irradiation 114, 117, 120, 121, 152, 156, 157, 160, 162, 163, 177
  warming 9, 22, 23, 29, 31, 41, 55, 262
greenhouse 3, 6, 14, 15, 17, 23, 28, 29, 31, 32, 36, 255, 260
  effect 22, 36, 42, 69

H

heat transfer coefficient 243, 244
homoscedasticity 119
hour angle 78–80, 88–90, 92–97
hydro-electric energy 37
hydrogen energy 6, 41, 43–45, 255, 260, 264
hydrological cycle 33, 37, 241
hydropower 4, 35–39
  small 39
hydrosphere 14, 21, 26, 31, 32

Index

I

independence   110, 119, 189
industrial   6, 18, 30
  revolution   3, 10
insolation   28, 34, 37, 60, 67, 68, 88, 97, 103
inundation   16
inverse
  distance   209, 214, 228
irradiance   103
irradiation
  change   120

K

Kriging   122, 172, 212–214, 220

L

linear interpolation   212, 213

M

mean
  absolute deviation   111
  bias error   111
measurement error   103, 105, 106, 119, 164
methane   15, 23, 28, 30, 41
Mie scattering   70
mitigation   12, 18, 33, 39, 40
model validation   107
monochromatic radiation   64
monthly average daily   88, 113, 198, 199, 201, 202
  horizontal extraterrestrial radiation   113
  radiation   113

N

natural
  gas   11, 255, 259, 260
  resource management   13
near outlier   112, 113
nitrogen cycle   26, 27
nitrous oxide   15, 23, 31
non-renewable   10, 16, 17, 34
non-sustainability   10
normality   118, 127

O

obliquity   57
oil
  crisis   15

equivalent ton   7, 43
  spillages   14, 42
optical mass   64, 65
optimal interpolation   212
orbit   50, 57
  elliptical   59
outliers   112, 113
oxygen cycle   27
ozone   13–16, 26, 29, 31, 32, 42, 63, 70, 240, 260

P

parabolic monthly irradiation model   196, 201
particulate matter   33, 141
passive   241
photo-optical transmission   253
photosphere   48
photosynthesis   21, 26–28, 30, 37, 39, 40, 48, 51, 56
photovoltaic   37, 42, 43, 51, 65, 67, 102, 245, 247, 256–258
plasma   49
pollution   4, 10, 14, 15, 22, 29, 36, 38, 42, 262
polynomial model   156, 203
population   6–9, 12, 36, 51, 112, 186, 187, 262
  growth   1, 6, 7
positive feedback   32
precession   58
precipitable water   116
production   35
pyranometer   68, 87, 102, 104–106, 196

Q

quadratic model   153, 154, 206

R

radiant-flux density   51
radiation   177
  change   24, 25, 120, 164, 197, 198
  electromagnetic   37, 50
  flux   69, 89
  regional change   24
radius of influence   210, 214, 215, 223–226
Rayleigh scattering   62, 63, 70
reflected radiation   62, 63, 104, 105
regional
  covariance   216
  dependence   211, 216

variability 210, 214, 216, 217, 232
relative humidity 118, 140, 168
renewable 1–4, 10, 17, 257
   energy 10–14, 16, 21, 24, 35–38, 239, 259–262
robotic 1, 2, 177
rotational velocity 77

S

sea-level rise 9
shade-ring correction 105, 106
shortwave radiation 62
silica gel beds 263
simple power model 156, 269
societal disruptions 10
solar
   altitude 75–77
   beam 57, 60, 62, 63, 85, 91, 93, 251, 263
   cells 256, 258
   collector 202, 241, 242, 245, 246, 249, 261, 263
   constant 66–69, 89, 91
   energy 1–4, 12, 14, 15, 17, 22–24, 35–37, 39, 56–66, 74, 101, 117, 151, 202, 261
   home 260
   spectrum 53
   transmission 254, 259
   fusion 49
   geometry 74, 77, 78
   heating 241–251
      active 241
      passive 241, 261
   hydrogen energy 6, 41–43, 255
      system 43
   irradiance 53, 67, 77, 81, 87, 105, 141, 163, 182, 213
      polygon model 160
   irradiation 162, 168, 172, 174, 192
   photon energy 43
   power 36, 41, 51, 57, 67, 103, 252, 256, 258
   radiation 246
   system 18, 48, 50
   time 72–74, 91, 92
   transmittance 71
solarimeters 102
solstice 75, 76, 78, 81, 94, 141, 164
spatial
   correlation 210–212, 216
   dependence 209
   function 209, 217
spherical coordinate 79, 80
standard
   atmosphere 71
   time 72–74

Stefan's
   constant 244
   law 244
stratospheric 15, 16
stratospheric ozone 16
Student's t-test statistic 110
sunrise 75, 77, 88
sunset 75, 77, 93
sunshine
   duration 37, 88, 101–103, 108, 109, 113–121, 156
   recorder 102, 103, 116
sustainable
   alternatives 10
   development 15, 37, 39
   energy 37, 60, 239
   future 15
   society 18

T

temperature gradient 50
terrestrial irradiation 37, 89, 91, 116
thermal conductivity 242, 243
thermonuclear fusion 50
thermo-siphon 246
tilt angle 81, 82, 252
triple solar irradiation model 168
turbidity 70, 116, 151
turbulence 117, 151, 156

U

unrestricted model 118, 119, 126
urban heat island 164
U-value 245, 261

W

water
   energy 33
   resources 14, 33, 38, 39, 172, 264
   vapor 9, 21, 26, 31, 32, 53, 55, 62, 65, 141
wave
   climate 40
   energy 40
weather 3–5, 9, 16, 22, 25, 26, 30, 47, 56, 66, 75, 136, 137, 141, 159, 161, 162
weighting function 209, 214–218
wind energy 17, 37, 42
wobble 57–59

Z

Z-scores 172
zenith angle 60, 65, 71, 76, 85, 87, 89

Breinigsville, PA USA
27 February 2011
256439BV00006B/23/P